Rag
Stick
&
Wire

GW00728818

Rag, Sticks & Wire is the story of how Australians took to the air with energy, humour, courage and pride, as airline pioneers, aircrew and passengers. It is also the story of how small airlines started to serve the outback in the 1920s, spread their wings and flew across and then far beyond Australia.

BILL BUNBURY is a producer with the ABC's Social History Unit and Co-ordinator for ABC Radio National and FM in Western Australia. His previous books include *Being Aboriginal*, written with fellow ABC producer Ros Bowden, and *Reading Labels on Jam Tins—Living Through Difficult Times*.

Rag Sticks & Wire

AUSTRALIANS TAKING TO THE AIR

BILL BUNBURY

an
ABC
BOOK

This book is dedicated to the memory of two
remarkable airmen, Paul McGinness and
Charles Ulm, whose role in the Australian
aviation story should never be forgotten.

Published by ABC Books for the
AUSTRALIAN BROADCASTING CORPORATION
GPO Box 9994 Sydney NSW 2001

Copyright © Bill Bunbury, 1993

First published 1993
Reprinted April 1994

All rights reserved. No part of this publication may
be reproduced, stored in a retrieval system or
transmitted in any form or by any means electronic,
mechanical, photocopying, recording or otherwise,
without the prior written permission of the
Australian Broadcasting Corporation

National Library of Australia
Cataloguing-in-Publication entry
Bunbury, Bill
   Rag, sticks and wire: Australians taking to the air.
   ISBN 0 7333 0273 4.

   1. Aeronautics—Australia—History. 2. Aircraft industry—
   Australia—History. 3. Air pilots—Australia—History.
   4. Aircraft industry workers—Australia—History.
   I. Australian Broadcasting Corporation. II. Title. III. Title:
   Rag, sticks and wire (Radio program).
387. 70994

Designed by Geoff Morrison
Set in Goudy Old Stle by Clinton Ellicott, Adelaide
Printed and bound in Australia by APG, Maryborough

3-2-1695
9 8 7 6 5 4 3 2

Contents

Acknowledgments

My special thanks to those who helped me make the original radio programs on which this book is based. They include John White formerly of QANTAS; engineers George Roberts and Ernest Aldis; John Ulm and John Gunn for their assistance with this book and contribution to the original radio material.

I'd also like to acknowledge, in this context, the help of ABC Archives and Ann Thompson, Carl Akers and Edo Brands, the Radio National production team, QANTAS Archives and Barry Cullen. Joanna Sassoon and the JS Battye Library, Perth, assisted with photographic research and, for their computer and transcription assistance, I thank Pam Fraser, Sheana Masterton and my daughter Alison. I must also thank my wife, Jenny, for patient proofreading and advice.

Then there are the many contributors who gave generously of their time for the original radio interviews. Without them neither the radio programs nor this book would have been possible. This list includes Pauline Cottrill, daughter of QANTAS co-founder Paul McGinness, who is now writing her father's biography.

Lastly, I am indebted to those who supplied personal photographs for this publication: Frank Colquhoun, now planning his own photo history of Western Australian aviation, Nancy Button of Longreach, Ernest Aldis, Pat Jordan, Margaret McDonald, Scotty Allan, and Daughne Kelpe.

Introduction

In 1986, on the fiftieth anniversary of the much-loved DC 3, Gooney-bird, C 47, Dakota, call it what you will, I put to air a collection of interviews from folk who'd flown or worked on this famous aircraft. One of these, as I recall, was ex-RAAF wartime pilot Sturdee Jordan who crash-landed a cargo-carrying DC 3 in northern New South Wales. He survived unhurt. So did the freight, 30 000 eggs, and not a yolk broken.

Public response to these tales encouraged me to do more. I sensed a strong interest among the audience in stories about flying. I went to see one of Australia's most distinguished early aviators, the late Sir Norman Brearley, and found a man in his nineties with an undiminished passion for flying and a sharp memory of his own efforts to get into the Australian skies after World War One. Taking his pioneer efforts as a base for what followed, I broadcast, on ABC Radio National, a four-part documentary series called *Flying out of the West*, a social rather than a technical history of aviation in Western Australia. The response was gratifying, but one appreciative call made me realise I had only scratched the surface. A staff member of the National Library in Canberra told me how much he had enjoyed the series and added that the National Library took more enquiries about aviation history than any other topic.

Within a year I had completed *Rag, Sticks & Wire*, an eight-part radio series which looked at further aspects of civil aviation in Australia and, in particular, the development of QANTAS from a small outback airline to its present status as a highly respected international operator.

Wherever I recorded passengers, cabin stewards, air hostesses, pilots and engineers, I also found delight. They were, without exception, men and women who had had a lot of fun and adventure out of aviation and were proud to have been part of its development. And it has been a remarkable part of

Australia's history. There are people still around, like Edwin Chapman and Mary Yeoman, who can remember 'rag sticks and wire' aeroplanes and can still recapture the thrill of their first bumpy and unnerving flight.

This book combines material from *Flying out of the West* and *Rag, Sticks & Wire*, giving a regional as well as a national focus to the story of civil aviation. Again the emphasis is social rather than technical or political. John Gunn's excellent three-part history of QANTAS 1919–1970 already provides a broad picture of national civil aviation. But readers who hope to find the history of Australia's many small airlines in *Rag, Sticks & Wire* may be disappointed.

I am well aware of the omissions but believe that the history of each airline is worthy of treatment in its own right, treatment well beyond the scope of this book.

Until such accounts are written, simply adjust your seat for take-off and enjoy *Rag, Sticks & Wire*, a sample of personal histories of Australian aviation.

Bill Bunbury
Perth

Getting off the Ground

At that time I had acquired my first motor-cycle and as I rode this motor-cycle round sharp bends I realised I was leaning over at quite a steep angle. And I said to myself, 'Just like aeroplanes have to do. That's it! I'm flying! I can fly! Of course I can fly, I've just got to do what I'm doing on the motor-cycle. So I can say I'm in the business of aviation because I've just got the urge to fly!'

Norman Brearley, founder of Western Australian Airways and inaugurator of Australia's first air mail service.

Aeroplanes those days didn't take anything like the long take-off run and landing run that they do now, and there were no concrete runways. They just had to land and hopefully the undercarriage would take it. They were rag, sticks and wire, well, I mean, they were just the beginning.

Edwin Chapman, central Queenslander and early aviation enthusiast.

THE CRITICAL YEARS 1919–1921

*T*he story of how Australians took to the air early this century is a tale of adventure, risk, obstinacy and ingenuity in the face of indifference and setback, a story remembered by people living now, men and women who not only took part in it but shaped it, a story with all the qualities of myth and legend but with very practical beginnings.

Above all, taking to the air was essential if Australia was to bridge its vast spaces more efficiently. Rail transport had begun to dent distance but terrain and climate worked against rail, especially in remote areas. Floods could wash away the best ballast and the heat of the Nullarbor plain in summer frequently warped the rails of the brand new Trans-Continental line between Adelaide and Perth. In eastern Australia the Murray was still a major transport artery and most capitals relied on sea connections rather than those of road or rail.

When civil aviation emerged in Australia as a new partner in transport it was to have more than one birthplace. While aircraft would eventually serve the major population centres of the south eastern seaboard, they were before long an indispensable part of the life of remote northern townships—Winton, Longreach and Cloncurry in Queensland; Onslow, Derby and Broome, in north west Western Australia. Fly to any of those towns today and you're still conscious of their isolation.

I recall sitting up front in the turbo-prop which took me to Longreach to record interviews for *Rag, Sticks and Wire*. We had been flying for two hours out of Brisbane. At 20 000 feet the bleached landscape of central Queensland shows little obvious sign of life. The pilot pointed out of the left hand window, 'That's Longreach'. It was a dark brown dot in the huge burnt plain below and still half an hour away. It was at that moment that I fully appreciated just what the early aviators brought to the outback, men who had only recently soared above the pervasive mud and grey cold of the battlefields of northern France and Belgium.

A photograph from the time shows several allied airmen at the wartime funeral of their respected adversary, the air ace Baron von Richtofen, shot down over the Somme in 1918, pilots who were soon to make their mark in the world of Australian civil aviation after the war: men such as Horrie Miller, Charles Kingsford Smith, Hudson Fysh, Norman Brearley and Charles Ulm. All were luckier than the infantrymen below them, soldiers whose chance of survival grew less with each passing day on the western front.

Yet casualties in the air were bad enough. Norman Brearley, who went on to found Western Australian Airways in the early 1920s, was soon convinced that pilot survival depended on proper training. His own injuries gave him ample time for reflection.

> Midway through the first World War, one man in France saw the terrible effects of not having trained pilots to go to war and he caused quite a stir with higher authority by saying, 'You must have trained instructors to teach pilots how to fly.' After a lot of argument he won the day and Great Britain set up what were called Schools of Special Flying. I had the good fortune, after I had been shot down with a bullet in both lungs, to be stationed at one of those schools, which enabled me to teach instructors how to instruct.

Hudson Fysh also began his aviation career in World War One.

> I owe my own start in flying to the first World War. I went away with the Light Horse and out at Gallipoli we went through some fairly tough times. I was in Palestine with the Light Horse and I had some friends over in No 1 Squadron Australian Flying Corps, and I thought, well, it would be a fine thing to learn to fly and join these people. Of course, when they chose a pilot in those early days, they used to ask you, 'Are you a horseman? Have you good hands with a horse?' It was a personal thing. You flew with your hands and the planes could be pretty tricky at times. If you didn't do a turn in the correct way, I remember, certain types used to go into a spin which used to take you a thousand feet to get out of.

It was probably no coincidence that many of the famous names of early Australian aviation all saw active service in the Royal Flying Corps in World War One.

If they could survive the hazards of military action and the unreliability of their aircraft, they were more than ready for the risks and the opportunities of civil aviation when the 'war to end all wars' finished in November 1918. Their skill and experience in the air had equipped them to try for a living in a new and risky industry, and in Australia the pioneers of aviation challenged distance and isolation on both sides of the continent within two years of the end of the war.

And what a challenge! It wasn't just setting up airstrips and supply bases, getting hold of suitable aircraft and qualified staff, all difficult enough when there was little expertise around and little knowledge amongst the public of what was needed. But the hardest task was persuading potential passengers that aircraft were practical and safe. And, curiously, aviation pioneers taught people to take to the air by barnstorming—thrilling exhibitions of aircraft and pilot performance with all the risks and sometimes tragic consequences.

Nevertheless these exhibitions attracted huge crowds and sometimes passengers. In Western Australia, country schoolgirl Gwynneth Ayers was one of the original enthusiasts.

When Norman Brearley came back from the war, in 1919 I suppose it would be, he gave this aerobatical display. And, as you can imagine, all and sundry went along, as many as could, to what is now Gloucester Park for this display; and it was really fascinating.

I had never seen an aeroplane before, being a country child; it was all new to me, and to see this plane coming down as a falling leaf, just gently gliding down to the ground. And then there was looping the loop! It was really wonderful.

Well, after a while, unfortunately, the plane contacted one of the wires around the grounds and came down. Nobody was hurt but poor Norman Brearley was very distressed, of course. The afternoon was spoilt for a time. But then he decided he'd rush off to Maylands where he had a second plane. And they jumped in a car and went off. All the crowd stayed, of course! And I suppose it would only be a matter of twenty minutes and he was back on the grounds to carry on with his display!

Well, during the afternoon he gave flights to everybody— I believe there was a charge—and the one who impressed me most was the one and only woman that ventured up.

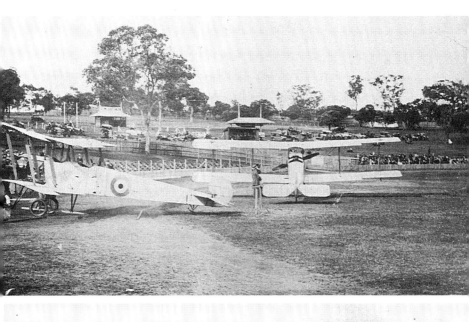

Top: Norman Brearley flew these planes in aerial displays in 1919
Bottom: Norman Brearley at Harvey, barnstorming in 1921

That was Miss Violet Peet. We thought she was very brave, but she went up and she came down all in one piece.

Miss Peet and other Perth enthusiasts for the new wonder of aviation saw aircraft very much as a novel form of entertainment, perhaps a city perspective on technology that seemed to belong more to the circus world, to stunts rather than schedules.

Norman Brearley was fortunate in soon finding a patron who took aviation seriously. Michael Durack, State Member of Parliament for the Kimberley, needed no persuasion of the potential of aviation. It took him ten days to travel by steamship between Perth and his remote north west electorate. By air he could make the same journey in two days. His daughter, author Dame Mary Durack, recalls how he gave Brearley his much-needed start.

> My father suggested that he could have part of our backyard which ran down to the Esplanade, as we called it then. We lived at the time on the corner of Victoria Avenue and Adelaide Terrace, 263 Adelaide Terrace, which is now a high rise Perth building. He could have his hangar in our backyard only if he could get permission to use the Esplanade as a runway, which my father, being then in Parliament, seemed to have had no trouble in getting. To their mutual delight it was sanctioned and he set up his aerodrome at the back part of our garden, and there were his planes and his mechanic who used to bring us boiled lollies! We must have been a terrible nuisance. We were fascinated with what was going on down at the back. But my father said he wouldn't charge him any rent for it but he wanted one thing and that was to be with him on the first official flight.

•

On the other side of Australia, returned airman Hudson Fysh was also about to convince inland Queenslanders that aircraft were not merely a form of entertainment. His chance came with the setting up of the England to Australia air race in 1919.

> I had my partner with me, Paul McGinness. We were both twenty-three years old. He was a pilot and I flew with him very often as an observer before I got my wings. I remember on the troopship, just before we got back to Tasmania, the

Paul McGinness (left) and Hudson Fysh in World War One

news was flashed through that the Commonwealth government had offered a prize of £10 000 for the first Australians
to fly from England to Australia.

Unfortunately for Hudson Fysh and McGinness, their attempt
to enter the race ended with the unexpected death of their
sponsor, Sir Samuel McCaughey, a rich banker and grazier from
New South Wales. He had been won over by McGinness's
enthusiasm, but with his passing the pair had to content themselves with a Commonwealth government offer to survey the
route the fliers would take over north east Australia. The trip
turned out to be more than a consolation prize for Fysh and
his partner:

> Going through on this 1919 survey, we met Sir Fergus
> McMaster and other Queensland graziers and squatters. We
> found that they had a great problem out there. They had no
> main roads, no bridges over the rivers, which only used to
> run, of course, once or twice a year. But when they did run,
> everything came to a standstill. And they thought, now this
> is just the place for an aerial service. They were keen. We
> wanted a job. We were young and energetic. So we formed
> QANTAS with these people.

Not ordinary people either. Fergus McMaster was a hard-
headed Queensland grazier. But a chance meeting with Paul
McGinness in Cloncurry set the scene for the beginnings
of the Queensland and Northern Territory Aerial Service.
McMaster's car had broken down, McGinness fixed it and the
outback pastoralist and the energetic young aviator became
friends, a friendship that developed into partnership in 1920
when McMaster and his fellow graziers found the capital to get
QANTAS into the air.

Paul McGinness today is the forgotten man in the QANTAS
story, but those who knew him as a young man sensed both
charm and drive. Longreach resident Nancy Button took her
first aeroplane ride as his passenger at the age of six.

> He used to come out to 'Dundee', my father's property,
> which is about six miles east of Longreach. Why I remember
> it is because we always had duck to eat after Paul McGinness

had been. He would fly around the waterholes on 'Dundee' and spot where the ducks were. He would then ring my father and he'd fly out and land at the back of the house at 'Dundee' on a ridge there, and then they would go up duck-shooting; and that is really the very first that I can remember with McGinness.

One day he came out, and while he was at 'Dundee' a ring came from Longreach to say that this present airport had been graded, and that he could land on that. They rang from town. So I can remember my mother dressing me up in a black woollen overcoat and goggles, because the planes were open to the elements, and I flew in with McGinness.

Well, you can easily tell by that that he took trouble with children. If he'd have been bombastic or anything like that, I'm sure he wouldn't have even thought to bring the little girl from 'Dundee' in to Longreach just for a ride.

Sadly, McGinness didn't stay to enjoy the eventual success of the airline he pioneered. He was a very different man from the more dogged Hudson Fysh with whom, it seems, he had early and serious differences of opinion. One concerned pilots and drinking. Fysh and other members of the QANTAS board sought a pledge of temperance from their flying staff. McGinness felt they were being unrealistic. He recommended instead that pilots be urged to drink moderately and well away from their flying duties. The Board did not agree.

McGinness left QANTAS and Queensland early in 1923 to pioneer virgin country, growing wheat and sheep in the Morawa district of Western Australia. His marriage failed and the Depression of the 1930s took its toll of his farm. His health was also beginning to fail. But in World War Two he joined the RAAF, becoming a Squadron Leader. He trained pilots at Parkes in New South Wales and later served in New Guinea. After 1945 he tried growing tobacco in northern Queensland without success. On medical advice he left the tropics and again tried growing tobacco in Northcliffe in south western Australia. He died in January, 1952, aged fifty-five, and was buried in an almost empty plot at Karrakatta cemetery, Perth, with no recognition of his service to both military and civilian aviation. The only people to attend his funeral were the priest, his first wife and his younger daughter Pauline.

QANTAS honoured his memory a year or so later, placing a bronze plaque on his tombstone.

Right from the beginning McGinness had believed in the potential of aviation, an industry he had described during World War One as being only in its infancy. Speaking after the QANTAS inaugural air mail flight on November 22, 1922 between Charleville and Longreach, McGinness said that he saw

> people using aeroplanes like trains and buses. This company will one day fly all over Australia. Indeed one day we will be flying around the globe.

McGinness's partner, Hudson Fysh, was not to retire until 1975. In temperament quite different from the mercurial Paul McGinness, he dedicated himself single-mindedly to the new technology. As his son John has come to see it, Hudson Fysh's commitment to aviation was quasi-religious.

> An aspect that has passed me by, and certainly his biographers, has been his evangelical approach to aviation. Whilst he wasn't preaching religion, he was preaching aviation because after QANTAS got started, it had to be proved to people that air travel was safe and reliable.
>
> For a man who wasn't given to public speaking he set about doing just that, and he had put together a number of slides (the old glass slides about two inches square) and an old machine, and he went about the country talking to Rotary Clubs, any club at all.

But proving the worth of QANTAS was easier said than done. Like Brearley in the West, Hudson Fysh and McGinness still had to sell aviation to the public. While hard-headed squatters like Fergus McMaster could see the obvious advantages of aircraft in difficult country, it was the centres of population that would provide passengers. Accordingly, the fliers barnstormed their way through the back country, attracting adventurous spirits like Mary Yeoman, one of Hudson Fysh's earliest passengers.

He went all through the back part of Queensland and that's

Top: The little girl in flying goggles is six-year-old Nancy Button, pictured with her grandparents on 'Dundee' after her flight with Paul McGinness. Bottom: McGinness photographed in the late 1920s on his farming property after he left QANTAS.

where your QANTAS comes in—Queensland and Northern Territory Aerial Service. Then he went all through the area, all through the district, and took joy flights—three guineas for fifteen minutes. We were getting five shillings a week working in a shop in those days. Of course, everybody flocked out to each little country town he went to because there was very little entertainment then. A lot of them wouldn't go up, but I said, 'Oh no, I'm going up. That's what we came out for.' We had all our rides and I was sitting there—and no security. There was no wall around you or anything in those days. You just sat there. You could put your elbow out the window like a train. And I'm sitting there with the goggles (you had to have a helmet and goggles and a mask) and all strapped in like a kiddy's safety harness. Anyhow we're sitting there and sitting there and his assistant's down on the ground, and Hudson Fysh said, 'Contact!' to him, you see, and I'd been sitting there for some time, and I thought he said it wouldn't act.

I thought, how am I going to get out of this? And I thought, oh goodness! So I shut my eyes and when I opened them I was up in the air. I looked down and I thought, oh well, I'll come down somehow. But we had a lovely flight around and I did thoroughly enjoy it and, of course, he said to everybody, 'How did you enjoy it?' And I said, 'Oh, I loved it!' I said, 'If I ever live long enough and have enough money,' I said, 'I'd love to fly round the world'.

Mary Yeoman did fly round the world some fifty years later, in the comfort of a four-engine jetliner. And when she described her very first flight to a QANTAS hostess, she was invited to sit up front with the driver.

Edwin Chapman of Longreach was also one of those early adventurers.

I had a flight with Hudson Fysh when he was actually floating the company, flying around canvassing shares and giving joy-rides and showing what flying was like. The vibration was absolutely terrible. I was watching the side of the cockpit and then oil started to come back and over my goggles. I had to pull my handkerchief out and keep wiping the goggles and I got some of it in my mouth. We flew around for ten minutes and, of course, it was all fascinating to see these little houses underneath, like little dolls' houses, and one thing I saw was the mist on the horizon. And that was something I hadn't seen before either. Anyway, when we landed I felt as though ... oh gee! I felt sick. It was all the burnt castor oil in my mouth.

Despite doses of such unwanted medicine the post-war generation of the 1920s needed little exhortation about the practical benefits of aircraft. For them the aeroplane offered much more than simply thrills and spills. It was country Edwin Chapman grew up in and knew well.

Between here and Charleville it was virtually impossible to travel in wet weather. If it rained at all and there were creeks that were up, if the river was running or anything like that, they just couldn't travel. You know the aeroplane was the only means of getting from Charleville to Longreach, and from Longreach to Cloncurry, because there were no roads.

If you got outside the outer suburbs of Brisbane that was the end of the roads. You were in the bush, and that's how it was out here—just bush tracks. So that the start of QANTAS made a tremendous difference to communication.

Hudson Fysh and McGinness aimed to link three towns, Winton, Cloncurry and Longreach. All were sizeable centres and each was served by a railhead running in from the coast. It was the country between them that took time to travel—hot, dusty and corrugated in the heat of summer, or isolated by flood in the wet. Even if they couldn't entice passengers into the air there was always the mail.

A letter from Brisbane might come by train as far as Cloncurry, wait unopened for weeks while floods slowly went down and then travel by buggy to an outlying property. Both families and businesses stood to gain from the benefits of air mail, an advantage pressed by both Brearley in Western Australia and Fysh in central Queensland.

Jack Hazlett was one of QANTAS's very first engineers. He felt that in those early risky days it was the government air mail subsidy of three shillings and sixpence which kept the airline going.

We didn't get many passengers because, you know, the publicity wasn't too good. Mostly they were damned glad to get off the plane; felt sick even if they didn't vomit, sat in a hard little seat. There were no comforts there and the noise and the oil mist flying past there, even with cap and goggles on it would be dribbling down. It was, I think, a very sturdy person who'd want to use the service.

Not surprisingly there were still doubters. Audrey Mcleod's father was one Longreach businessman approached by Hudson Fysh.

At that time he was starting the QANTAS, you know, Queensland and Northern Territory Aerial Service, and he was naturally looking for money and he tried to interest my father in putting money into the business. But Dad wasn't having any. Dad was one of the old school who felt that he should keep his feet on the ground, and he didn't follow up with Hudson Fysh … much to his later disappointment.

One who did follow up was veteran grazier Alexander Kennedy. He had already explored much of northern Queensland in the 1890s using bullock teams, losing his partner in the process to spear attack by Aboriginal warriors. Over thirty years later and aged eighty-four, he was a QANTAS shareholder and passenger, climbing into an open cockpit and shouting, 'Damn the Doubters', as the plane taxied towards take-off.

Aviation not only brought a strange new technology to the outback. It also brought a different sort of person into contact with grazier society. Hudson Fysh's daughter, Wendy Myles, recalls local reaction to the arrival of her parents.

I think they were regarded as something very different. Nobody could put them in a pigeonhole because my father had come from a very—I think 'refined' home is perhaps the way I'd say it—and so had my mother. So I think they found it very difficult. Certainly my mother did, and I think, you know, when my father used to fly down and land and sort of ask for money to start off QANTAS, it was a bit of a novelty, but quite different when he brought a wife.

My mother was really quite shocked, I think, by life in Longreach when she first went up there, and I don't think to her dying day she really ever forgot just what it was like going up there as a young bride. My brother didn't see rain until he was four and one of my family's stories is of my brother's terror when he saw it rain. You know if you haven't seen rain until you're four it's quite something. And, of course, there was no fresh milk of any sort, no fresh vegetables, no fresh, really fresh, food and I think it must have been very difficult for her with two small children.

John Fysh knew what it cost his mother.

She gave up a life in New Zealand. Although she was born in Rockhampton, the family had moved to New Zealand and she was brought up there. Well, the change from New Zealand was quite remarkable. There she was, in a little weatherboard cottage with a tin roof and no insulation, bore water for the shower, and open top tank out the back which she used to immerse herself in on the hot days, limited rainwater and no intellectual compensations. She was one of the few people that read, she and my father … he was a great reader. They had a library and the few people there, some of the pilots, like Charles Scott, used to borrow their books. But it was a very hard life for her and she didn't leave Longreach in seven years.

She was a very loyal person and she did a tremendous amount to help him personally, because she was ambitious for him and for herself, and her support was enormous.

There are a number of well-known stories, one of which was how she'd go out into the airport and dig her high heels into the ground to see how soft it was, and its suitability for landing aircraft. And she used to prepare sandwiches and things for passengers and I suppose she was almost the first QANTAS caterer.

Most people with an interest in aviation history believe that QANTAS was the first Australian airline to fly scheduled routes, starting as it did in November 1922. As far as passengers go that's correct, but Norman Brearley's Western Australian Airways flew the first scheduled air mail service in Australia between Geraldton and Derby.

It was a flight that would end in tragedy but not defeat for Brearley. He had prepared as carefully as possible for the big venture, taking particular care in recruiting his flying staff.

One was Charles Kingsford Smith, who undoubtedly was one of the finest pilots I ever came across, and another was Len Taplin, who was also a very clever and expert engineer. With those two fellows we set off on the first air mail contract flight.

The inaugural air mail flight could not start from Perth but had to begin at Geraldton, 300 miles north of the state capital. This was still the age of steam. The mail could only be carried to Geraldton by the Midland Railway Company. Air mail services, if they were to succeed, would have to challenge the rail monopoly. Ten years later, when aviation had begun to prove itself, the problem still persisted. The newly formed partnership of Imperial Airways and QANTAS could not fly letters between London and Sydney. They had to drop their bundles, as it were, at the railhead of Cootamundra, well up-country from the preferred destination.

So on December 5, 1921, when Brearley's trio of Bristol Tourers left Perth for Geraldton, their run was essentially a positioning flight. Frank Colquhoun, later to become Brearley's apprentice mechanic, describes the plan for the flight of the three aircraft.

One was flown by Major Brearley. He had with him MP Durack; and the second one was flown by Len Taplin and he had with him a mechanic named Jack Trestrail. The third aircraft was piloted by Bob Fawcett and he had with him another mechanic named Ted Broad.

They got to Geraldton all right. The next day they set off for Carnarvon and the aeroplanes in the first day, I understood, were fairly close together and all within sight of each other. On the second day there was, I believe, a little tendency for them to spread a little. They weren't in any sort of close formation, but they were all within sight of each other.

Brearley was flying parallel with his fellow-pilots when

Taplin noticed his engine was losing a little bit of power and he decided he'd better go down and land and check over his engine, just north of Geraldton, before carrying on. I watched him land and I saw that he landed safely, but I stayed at about 3000 feet. Unfortunately, Fawcett went down to a low height and watched Taplin land and there, neglecting his own flying skill, he took his attention off his flying. The aeroplane stalled and it dived into the ground from a height from which he had no hope of recovery, and he and the mechanic were both killed.

Gloria Hazelby, the wife of the station-manager at nearby Murchison House station saw what happened from below.

Brearley had come in and signed to them to land there and for the other one to follow him. Instead of that they kept flying around just over the tops of the trees, low, too low! Turned on its nose and came down. One wing hit the tree and the plane came down on its nose. I never thought about them being dead. I thought they'd need hot water. I saw Brearley's plane go over and I said to the native women, 'Run up and see that there's plenty of hot water'. I had a couple of beds on the verandah ready for the injured men.

It was very hot and, of course, Taplin as he ran over to the plane, turned round and said, 'Put out any cigarettes!' I can remember that. I didn't talk to Brearley much. It was the first time ever I saw a man cry.

Brearley, when he got to Geraldton, Durack and Jacobi got in with him to come back, and they held the inquest that afternoon because it was a very, very hot day. They made the coffins and I gave them white sheets to wrap them in.

It was left to Gloria's husband, James Hazelby, to conduct the funeral service. Both men, pilot Bob Fawcett and his mechanic Edward Broad, were buried that day in hastily improvised coffins at Murchison House. So too, it seemed, were Brearley's plans.

Mary Durack Miller, whose father had backed Brearley's venture from the start, had seen the trio take off from Perth only two days previously.

It was a very sad affair. I can remember being down on the Esplanade with great excitement, waving goodbye to them all, and I can also remember the sadness and the quietness of the return of the plane. It was a setback, I have no doubt, but it didn't stop the plan.

Frank Colquhoun remembers the pressure on Norman Brearley to abandon the whole project.

Top: L to R: Charles Kingsford Smith, Robert Fawcett, Norman Brearley, Leonard Taplin, Val Abbott. Centre: The crashed Bristol at Murchison House Station. Bottom: The graves of EW Broad and RN Fawcett in Murchison House Station Cemetery.

But it was only his own determination and his confidence in what could be achieved despite the setback that I think overcame the opposition that was put up.

The Murchison House crash put Brearley out of the air for several months, time that Hudson Fysh continued to use building up his infant airline in central Queensland. Not that Brearley and Fysh, at this stage, were in any kind of competition. Brearley sent Fysh WAA airline schedules for comparison and offered the benefit of his own experience. That relaxed relationship would change in the early 1930s as the possibility of a national airline became apparent. But in 1921 both airlines were at the beginning of a critical period. The next ten years would make or break civil aviation in Australia.

Other Routes
Other Lands

It was the first sight of another land after leaving Australia … and even going out through the back country for the first time was an amazing experience. But to see this other land with all these other people come over the horizon … that was really something.

QANTAS pilot Lennie Grey

The mails were going by train, taking three and a half days longer than if they were taken by air. And as you know, the railway line between Kalgoorlie and Forrest is the longest straight in the world. On the Perth–Adelaide run we had our own headlight on the aircraft which would show up the railway line along which we travelled.

Sir Norman Brearley, founder of Western Australian Airways

1921–1935

*I*n the 1920s many Australian aviation companies were formed, rose briefly and fell. QANTAS and WAA were two of the survivors, although the name Western Australian Airways would also disappear by 1935. The next ten years would prove the most testing.

In Western Australia, the Murchison House tragedy of 1921 had given Norman Brearley grim confirmation of his own concern about the provision of emergency landing grounds. Engineer apprentice, Frank Colquhoun, who was to stay in aviation until 1969, joined Western Australian Airways in 1923. As he recalled, Brearley's biggest problem for some years was the preparation and provision of suitable landing fields,

> not only the main landing fields at the various towns and ports of call but emergency fields in between. These were all important to provide safe emergency landing sites. As a matter of fact we used to carry on the aeroplane a portable telephone, like a field telephone. And, as the air route substantially followed, or was within reasonable distance of the telegraph line, in an emergency pilots were able to walk to the nearest telegraph line, climb up a pole, hook on their wires, wind up the thing, send a signal and get a message through.

Brearley himself had great difficulty explaining to the local authorities along the north west route, just what was needed for safe landings.

> They had never seen an aeroplane in the sky before, were guided by sketches which were incomplete and made no reference to the landing speeds of aeroplanes or the length of run required or the effect of rain, because they never knew what was required of an aeroplane. The aerodromes selected by the Road Board (or some other authority) had stumps and wheel ruts that would make them perfectly unsafe for aeroplanes to land on.

At one landing ground, Wallal, between Carnarvon and Port Hedland, Brearley's agents had assumed that aircraft would descend vertically—at least, that was how they prepared the ground.

They were told when the aerodrome area was cleared and

ready for landing, to put a white circle of stones on the ground, flush with the ground, and pilots would recognise that as being a prepared aerodrome.

Unfortunately, the local people, instead of making it one foot wide, or two feet wide, made it two feet high as a wall, and cleared the inside of that hundred foot circle.

And in central Queensland some aircraft, as aviation historian John Gunn observed, didn't always turn up when expected. They'd made a forced landing somewhere on a station property.

QANTAS would ring the various stations and find that it had passed over a station at a particular time. They'd get the truck out and go and look and find it in a paddock, and they would put big wooden derricks up and Arthur Baird would sometimes completely change an engine in the middle of a paddock, and it would fly out again.

Engineer Arthur Baird had been Flight Sergeant with both Fysh and McGinness in World War One. He was now to give QANTAS twenty-eight years of engineering excellence.

Jack Hazlett, one of Hudson Fysh's first engineer recruits, remembers the fragility of civil aviation in those early days. QANTAS was a struggling small organisation with fewer than ten staff, and it was the prospect of getting mail quickly that kept the public on side, not the passenger service. Hardly surprising given their choice of aircraft.

We were trying to tune up the second-hand Armstrong Whitworth planes that had come out from England as a stop-gap. We were expecting a magnificent modern aeroplane that Vickers had designed and built, the Vickers Vulcan. But when it arrived it was a ghastly disappointment. None of us had seen a plane like it.

We were all used to open cockpit planes with caps and goggles, and there were these velvet chairs, up and down, six each side of this very spacious cabin. You could stand up in it and there were hat racks all along, above, to put your little parcels in. It was fascinating! But the thing failed. On the tests of the one machine they sent out, the thing was a ghastly failure.

It was under-powered, hot and badly ventilated inside, and they were never built for going into hot climates, and had to

Top: One of QANTAS's first offices at Duck Street, Longreach, in 1921
Bottom: Early flying days: Paul McGinness, the pilot, is third from the
left and mechanic Arthur Baird is fourth from right.

be taken back to England. That left us flat, still struggling with these other old things.

•

Resignations were not infrequent in those early years. Founder pilot Paul McGinness left early in 1923. Hazlett himself quit in 1925 to join a motoring business. But the airline struggled on. By 1926 it was not only flying aircraft, it was building them under licence to de Havilland. Hudson Fysh believed that was the company's most remarkable achievement.

Arthur Baird, our chief engineer, and his staff, built six DH 50s, four-passenger aeroplanes. I well remember that our first one, built in 1926, was tested by me one day, got its certificate of air-worthiness the next from the Civil Aviation Department. The next day it was christened by Lady Stonehaven, the wife of the Governor General. We then flew the Vice-Regal pair away on a long trip to north Australia in that very same plane.

And on the opposite coast Brearley finally got his scheduled air mail service going a few months after the Murchison House accident.

Later he was to turn his attention to linking Western Australia with the east. In 1928 he opened a service flying Hercules aircraft along the 2000 kilometre route from Perth to Adelaide.

What sort of passengers did the new service attract? Gordon Appleton was a sprinter who needed to get to Adelaide to take part in the 1934 South Australian athletics championships. No-one from Western Australia had ever tried himself against interstate competition. He wanted to combine holiday, athletic prowess and the excitement of flying.

Appleton had good cause to remember his trip to Adelaide. Once there he won both the 100 and 220 yard races in the SA track events. But equally exciting, for him,

was the flight from Forrest along the original part of the Trans-Australian line. It was rather exciting at Forrest because it was fairly late in the evening and by the time we reached Forrest it was dark, and I was able to experience a night landing which nowadays is most commonplace.

There was a ring of lights, not a great number of lights as I remember, but adequate, and the plane made a perfect landing.

On those Trans-Nullarbor flights the method of night-landing was that used during the First World War, kerosene-soaked rags hung out on a line into the wind enabling the pilot to see the wind direction and whether the area he'd selected was safe to land on.

Appleton's flight made the usual overnight stop at Forrest on the Nullarbor plain, alongside the Trans-Australian railway.

Forrest had hostel accommodation because night flying wasn't a common practice. I think we had a room each. It wasn't a very large building but there weren't many passengers either. We went to bed almost as soon as we arrived. It was getting late, and we also had to be up at six o'clock to take-off for Adelaide, and that was by South Australian time.

Another, earlier, airfarer was May Macaulay. She flew for adventure on January 30, 1930.

I hadn't been in a plane of any sort before and this was most exciting. The passengers assembled on the tarmac outside the sheds and waited for the plane. Everything was as light as possible. And then the engines! At the take off we were still shocked with the noise.

She was even less impressed with the cabin service.

At Ceduna we had curried sausages and stale Swiss roll with powdered custard over it, and all of this was eaten at trestle tables with the wind blowing dust everywhere. It was a fairly open shed and while we were eating the plane was refuelled from a truck.

May has often wondered since why she went by air. Certainly her family thought it extravagant at the time. But it was something she wanted to do. In her own words she was too interested to be afraid of the consequences. She had rushed to the schoolroom window to see her first plane, piloted by Norman Brearley, in 1919 and had been bitten by the flying bug.

Brearley had always operated from a capital city, using Perth as his base, even if his initial aim had been to service

Top: A Hercules bi-plane at Maylands in the 1920s. Bottom: Ready to fly to Ceduna in 1932 to pick up a disabled engine

Australia's remote north-west. His choice, in linking Perth and Adelaide, served him well. He was now operating the largest air company in Australia and flying the longest route.

By 1930 QANTAS had also grown, thanks partly to government defence subsidies, but also through its own efforts in opening up the back country, extending, in Hudson Fysh's words, from Charleville to Cloncurry, then out to Camooweal, up to Normanton and to Brisbane, and across the Northern Territory to Darwin.

The growth in routes and traffic meant a shift for QANTAS from its base in central Queensland to coastal Brisbane. The airline had carried almost 10 000 passengers in its first ten years and had flown more than a million miles. Now, with the move to Brisbane, QANTAS was poised for expansion. As a boy, Hudson Fysh's son, John, was well aware of the significance of the move.

> Every weekend my father used to take me up to the airport, first at Eagle Farm and then later at Archerfield, and I can just remember a tremendous amount of energy being created. There were joy flights going on; even I was out in the car-park trying to sell tickets.
>
> The engineers were all there although it was weekends. Practically everyone was out there at the airfield doing something, and I remember this enormous energy and this feeling of progress and pride.

Lennie Grey had been determined to fly since his twelfth birthday. On that occasion his parents had given him a flight in a World War One bi-plane. In 1928 he had wagged school to see Charles Kingsford Smith land at Eagle Farm after his epic crossing of the Pacific. Leaving school he studied radio, acquired a commercial pilot's licence, flew unpaid with New England Airways and in 1936, at the age of twenty-one, joined the QANTAS staff.

> We had a very small office in 43 Creek Street, Brisbane. It had a ground floor and basement and a mezzanine floor. Now Hudson Fysh had the office in the back. The chief pilot was on the mezzanine floor and he had a desk, and the accountant, who became Sir Cedric Turner later on, he was down in the basement.

Cedric Turner was to stay with QANTAS until 1967, becoming General Manager in 1955.

Lennie Grey was also to remain with QANTAS until the advent of the 707 jetliner. But he spent his first flying years hedge-hopping with the mail.

Having to make so many stops, we couldn't go high even if we'd wanted to. We didn't have the aircraft to do it. Our first officer looked after all the mail, even starting from Brisbane, our terminal. The first job of the first officer was to go with old Arthur Baird, our chief engineer, who used to come in with his utility, and their first trip was up the back of the Post Office up the lane. And we physically went to the Post Office and collected the overseas mail which we took out to the airport, of course, and loaded it aboard the aircraft. But it was our responsibility right from the jump, every mail bag.

Fellow-flier Lew Ambrose also joined QANTAS at this time. Later he was to distinguish himself in wartime operations and became QANTAS London Manager in the 1950s, but in the 1930s he was also on the mail-runs, sometimes doing more than simply flying the aircraft.

We used to have to go to two western Queensland and Northern Territory cattle or sheep stations. We had a small hatch in the rear door, and we'd throw the mail out. On one occasion, as I threw it out, we hit a fairly heavy bump.

I took no notice and returned to the front, but when we landed at our next stopping place, it was drawn to my attention that there was some torn fabric on our tailplane, and the only conclusion I could come to was that in the bump the mail bag had been diverted and had probably struck the tail and torn the fabric. So I said to the chap receiving us, 'Could you mix me up some flour and water?' I'd tidied up the hole in the fabric and got two handkerchiefs out of my pockets, and when he got the flour and water I rubbed it all over the fabric and then laid the handkerchiefs tightly down and, of course, with the heat of the sun, they were dry almost at once and you could actually drum on them as though it was the same old fabric.

And, as Lennie Grey recalls, there was one occasion when the mail didn't arrive on time.

One of our number was a real wag. He took off from

Camooweal one day and, to explain it a little further, the mail locker was on the opposite side of the aircraft from the passenger door, so you walked right round and opened it and when you got everything in you locked it.

But on this occasion he couldn't have done so—not securely anyway—because after they took off he distributed the mail unknowingly right across the Northern Territory, and there were little mail bags everywhere, and even up to eighteen months later stockmen were coming in with these small mail bags. Of course they had the labels on so they knew where they were going. He got reprimanded rather severely for that little effort.

Catering was another matter altogether. No one went hungry on Lennie Grey's touch-downs.

We didn't cater for foodstuffs in the air at all, except in one section, I think. But our first stop out of Brisbane was Roma where our agent brought out morning tea, and that was set out on a table just at the side of the aeroplane, and our next stop was Charleville where there was a very good hotel. Old Harry Coroney out there owned the hotel, and he would bring out our lunch to the hangar, and Arthur Butler met us there from Cootamundra with the mail. Now it didn't matter to Harry whether it was middle of winter and freezing cold, or whether it was middle of summer and about 110° to 115° in the shade, Harry would bring out sizzling steaks. And we had our lunch there, and any through passengers, of course, had it in the old hangar.

Now on our next little catering jaunt we went to Charleville back to Longreach, and we'd pick up a few little afternoon bits and pieces, particularly cream buns. I don't know why we got cream buns, but they were very good cream buns, lots of cream sticking out, and I do remember opening the cockpit door (as it was then) and looking back at the passengers; and I happened to be eating my cream bun and they were all being violently ill. I don't know whether it was the sight of my cream bun that did it or not, but that did happen. That gave me extra work because I was the guy who had to get the bags out.

Kate Dean's family, from Longreach, often made the long trip down to the city.

I think it was in 1931 that Mother and my sister (who would have been very small at that stage, probably two or three) were flying to Brisbane. They used to have refreshments at their stops en route in QANTAS planes in those days, and Captain Russell Tapp was in charge, and my sister had been sick on the trip and to Mum's horror she saw the Captain stuffing her with cream cakes and lamingtons and whatever else had been provided for afternoon tea. She was horrified at the results that might occur, but he claimed that when people were sick the more they ate the better, and kept on filling the little girl up. It must have had some effect because Mum said she wasn't sick for the remainder of the trip.

Flying those journeys was still 'seat of the pants' stuff. Scotty Allan, who went on to become QANTAS Assistant General Manager, was well aware of the hazards for both passengers and crew.

Most of the passengers, unless they were old hands, got sick. I'm not surprised either. It was very bumpy. Worst of all in Queensland are dust storms and when I flew in Queensland you flew generally pretty close to the ground, bumps and all. The poor passengers had a bad time. Visibility would be fifty feet in the dust storms and to go from say Camooweal to Newcastle Waters you followed cattle tracks. Of course you had to know which cattle track to follow.

I was at Longreach one time and we could hear Captain Crowther, Willy Crowther (he's now dead), flying overhead but we couldn't see him. Arthur Baird was there, of course, at Longreach, and he said 'How about going up and fetch 'em down?' because he couldn't see where the aerodrome was.

I took off and went up through the dust and I found Willy all right, and Willy knew what I was there for and he followed me and I went down through the dust and landed, for which he was very grateful.

•

In the west Norman Brearley had enjoyed a clear run, with a monopoly on both the north west and the Perth–Adelaide run. But eventually his tender for the mail run was challenged by a South Australian–Victorian company, MacRobertson Miller

Airlines, in the person of Horrie Miller. Mary Durack, who was later to marry Horrie Miller, recalls what happened in 1934.

Brearley's time had run out and they were supposed to put in for another tender for the mail service. Well, I can remember my husband by that time was running an air service with MacRobertson in South Australia. My husband always remembers the day that he was tinkering as usual at something under the plane, when Norman Brearley emerged from an aircraft from Western Australia. He called out to him and said, 'I suppose you put your tender in,' and Horrie—we often told this—he went in and saw Bob Paterson as he usually did for lunch in Adelaide, and he said, 'I just saw Norman Brearley. He said, "I suppose you put your tender in." I don't know what he was talking about.' Bob Paterson said, 'Well, as a matter of fact, it had occurred to me you might be interested, but,' he said, 'it closes tomorrow.'

So the story was that they got straight on to the plane and flew from dawn until nine o'clock and got to Melbourne, and asked MacRobertson whether he was interested. He said, 'Well, if it's not too late we'll put in a tender.' So they did.

In Melbourne they'd assumed that there were no other tenders and that Brearley had the service again, but as it happened Miller's was the other tender, and for some reason they got that service. My father felt it was a shocking thing that Brearley would have lost that tender. He was always very supportive of Brearley.

Horrie Miller was a singular character in the Australian aviation story. He built and flew his own aeroplane before World War One and, like Kingsford Smith and Brearley, he flew in France with the Australian Flying Corps. Dame Mary Durack Miller remembers Horrie as an absolutely devoted aviator.

He just loved aeroplanes—had since he was a child. It had been the great romance of his life, the thought that man was one day going to fly. His father died when he was only fourteen and left him enough money to get him to England where he got a job in Sopwith's air factory, and then gradually got into flying himself. He was there when the war broke out; he put his age up so he could get in.

But they didn't want him. He and another couple of Australians applied for the Air Force and they said, 'No, we

Scotty Allan in RAF flying gear in 1918

want officer-types'. I don't think you could say that about my old man being an officer-type—at that time anyway.

So they came back and he was either the first or the second to register for the AIF. He went back and survived the war and was fighting in France with so many others; was there at the burial of Richtofen whom he could have been shot down by. It was a pity that great airmen such as these had to meet under unfortunate circumstances, instead of being able to meet on a friendly basis.

When Horrie Miller returned to Australia after the war he continued his flying career, even courting his future wife from the air.

I was in the north at the time and was going on the *Kalinda* to be bridesmaid to a girlfriend. Horrie for some reason or other was on the boat. He'd injured himself in some way. Something had happened and he was just going to take that little trip to keep quiet for a week or so, which he did. Well he happened to be sharing a cabin with my brother, Reg, who said he was 'a character of a chap. He was an airman but looked more like an old bushman. He'd got a swag and went on more like a bushman'. Anyway, Reg said, 'Would you like to come out to the station?'

If he ever came up after that he would do some circles and one of those stunts—falling plane stunts—to the great excitement mainly of the Aboriginals, who really thought he was going to crash. Occasionally he didn't land but he would throw something out and it was always a carefully wrapped box of fruit or something that we were very pleased about. So that developed as a friendly but fairly casual relationship.

Horrie Miller was to remain a dedicated aviator almost to the end of his life. Just after World War Two he bought an old pearling bungalow in Broome, at the hub of MMA's north-west operations, but far from its administrative centre. Impatient with office life in Perth, he was happier, spanner in hand, greeting the early morning planes in far-away Broome, checking aircraft and crew, even sending a pilot back once to find a missing engine cowling that had fallen off on the long flight north.

•

On the east coast QANTAS was on the verge of considerable change. The move to coastal Brisbane from inland Longreach had made QANTAS management acutely conscious of the possibility of routes overseas, creating links that were too difficult for a small airline to forge on its own, but with the right partner an exciting challenge, as Hudson Fysh's son, John, recalls.

My father could see that Imperial Airways had this knowledge of operating overseas airlines. He could see that the map between Australia and England was heavy with red ink and that, because of the British influence and the British association, QANTAS would get an enormous amount of help in establishing itself as an overseas airline.

In many ways the England–Australia link was part of the empire-building strategy. The world map of the time showed British possessions conveniently dotted all the way from London to Brisbane. The proposed new route would be the world's longest scheduled airline route, 20 000 kilometres.

But while Britain still saw itself in imperial terms it was important for the Australian airline to ensure that it got a fair share of the action.

QANTAS engineer Ernest Aldis saw aircraft evolve from Sopwith Camels to 747s. Having worked in the United Kingdom himself he was well aware of the English perspective.

The British Government used Imperial Airways as their flag carrier. They said if there were any flights to operate between Australia and UK, Imperial Airways was the airline which would get the nod. However, the Australian Government complained bitterly and said, 'This is not right. We want a slice of the action as well.' So finally Imperial Airways and the British Government decided to terminate their route at Singapore and give us a little slice of the operation—from Singapore to Australia.

Initially the Australian government saw its role in terms of completing the Imperial route, flying the last leg from Darwin down to Brisbane. But the government also had strategic objectives. The Defence department saw the link to Singapore as a means of ensuring Australian control of that route. Eventually it was agreed that Australia would fly to Singapore, and England would fly from England to Singapore.

Naturally the route was open to tender from the Commonwealth government and, at first glance, QANTAS, despite its experience of northern Australia, looked an unlikely contender with neither the aircraft nor the infrastructure to go it alone.

One possibility was partnership both with Imperial Airways itself and with other companies within Australia. An obvious partner, before his loss in 1934 of the Perth–Adelaide mail run, was Norman Brearley. When he first won that contract in 1928, he also won the longest air route in Australia, superseding QANTAS as the leading domestic airline in Australia. Brearley himself proposed a triple merger with Australian National Airways.

> We had meetings in Melbourne where we discussed the possibility of forming a company that we would manage and would operate such a service; and it was then, after various discussions, that Hudson Fysh suggested that the way was open for us to join forces with Imperial Airways.

•

Australian National Airways was the partnership of two of Australia's most famous airmen, Kingsford Smith and Ulm. Their overseas experience and reputation, with flights across the Pacific and to England, helped them launch ANA. Their first service ran between Sydney and Brisbane, using Avro 10 aeroplanes which could carry up to eight passengers. By contrast QANTAS had no multi-engine aeroplanes with which it could safely fly over water.

But the merger didn't come off, for various reasons. Brearley stated his:

> To me, this had no great appeal other than lending my knowledge and experience to those who would be interested, and the interest was created by Charles Ulm and Hudson Fysh getting into communication with me and saying, 'Can the three of us get together rather than try and go separately against each other on an England–Australia service?'

At this point one of the partners, Australian National Airways, faltered financially. Charles Ulm, who was the financial driving force behind ANA, saw that amalgamation just wasn't a realistic prospect. And, according to QANTAS, Norman Brearley wanted too much for the sale of his assets. Aviation historian John Gunn sums up the situation.

> They could see they would never get together and Ulm suggested that each take part of the route and do it independently but, of course, Ulm wanted the cream and he wanted to fly overseas, so did QANTAS, so did Brearley. It was competition not co-operation.

Brearley defended his decision to stay put.

> For my part I had no urge to leave Western Australia. I had plenty of strength here. I told the other two that I would only be interested in being a director of the organisation and providing a fairly large proportion of the capital that would be required. But again, I rather joined with Ulm in saying that I would not be interested in joining forces with Imperial Airways.
>
> Charles Ulm had no faith in Imperial Airways and he made it perfectly clear to Hudson Fysh that he would never be able to join forces with Imperial Airways. On the other hand, Hudson Fysh saw the strength of Ulm and realised that he would be overwhelmed by Ulm if they tried to join forces.
>
> So we broke up and Fysh threw out the bait to Imperial Airways who took it. And a cartoon of those days showed the Imperials at a table and the cartoon caption was, 'The Imperial meal', and on the plate was the skeleton of a fish.

Left on its own, QANTAS decided to link up with the British interest and, in partnership with Imperial Airways, form a new company, QANTAS Empire Airways. The QANTAS bid was, of course, still open to tender, with the distinct possibility that the very partners QANTAS had tried to woo might now be rival contenders for part of the London–Australia route.

John Gunn describes the final moves in this chess game.

> Eventually the battle for the contract for the air mail service came down to tendering, to lodging tenders on a certain date in Melbourne. Hudson Fysh and a chap called Dismore

from Imperial Airways, had spent a lot of work putting their tender documents together, and they put them in a black steel box and put it on the floor of their sleeping compartment on the train and went down to Melbourne; and who should knock on the door and come in to the compartment and sit down to have a yarn with them, but Brearley and also Ulm, and they put their feet on the tender box little knowing that all the QANTAS secrets were in that. Well, at the end of January in 1934, the tenders were duly lodged and QANTAS came out as the winners.

Hudson Fysh, of course, was delighted.

It was a very tight go but we managed to get it, and that was perhaps the greatest crisis that ever happened in the progress of QANTAS right through up to this day.

Just before Christmas, 1934, the Duke of Gloucester inaugurated the first Australia–England air mail service from Archerfield Aerodrome, Brisbane, with these words.

Each new method that makes for a speedier communication within the Empire must be of benefit to its people who are spread across the world. This mail plane is to carry your Christmas mail, and incidentally some of mine, to friends at home, and I commit them now to Captain Brain who is kindly doing duty for us as Father Christmas.

What QANTAS management and inaugural air mail pilot Captain Lester Brain knew, and the Duke of Gloucester didn't, was that the plane in question had no certificate of airworthiness. Lester Brain describes how they solved the problem:

Well, in these circumstances it was decided that we'd go ahead with the departure just the same. So we put our nice new DH 86 away in the hangar and shut the doors, and we got one of our previous big single engine B 861s out and gave it a coat of paint to make it look beautiful and fresh and with that backed up by another DH 50 the ceremony duly took place.

The Duke of Gloucester sent the first air mail off to England, and shook hands and so on; and I don't think that among the thousands of people there on that occasion to see the first Australia–England regular mail service start, even

half of them realised that it was the wrong plane going out—
that the real four-engine one was locked up in the hangar.

Heady days! QANTAS took possession of two new four-engine
DH 86 aircraft from England. Lester Brain delivered the first of
them. But the company's elation was short-lived. The second
plane, which was being delivered by Imperial Airways, crashed
near Longreach and all aboard were killed. Kevin Bower, then
nineteen, of Barsdale Station, and his friends were hoping
to catch a glimpse of the new QANTAS plane en route for
Brisbane. They saw it fly between the house and the station
dam when

> the thing started to roar like a car in a bog, and then all of a
> sudden it seemed to die and go into a spiral dive and we
> watched it, and then it went down behind the trees and up
> went the dust, and I said, 'My God, Andy, it's landed! We
> better jump in the car and get over!'
>
> The plane was crashed and a hell of a mess. We investi-
> gated and found two fellows hanging out the front and then
> we broke into the back of the plane and found another two
> fellows in the back, and we brought one chap up we thought
> was alive. He was quite warm, but when Andy put the mirror
> to his mouth we could see that he was dead.

There seemed to be no apparent reason for the tragedy. But
Lester Brain, with his own recent DH 86 flight from England in
mind, felt that the account of the crash reminded him only too
well of his own experience with the first aircraft.

> I flew out with Arthur Baird, our chief engineer, to the
> site and landed alongside it, and having heard the story of
> the gyrations that the plane went through, and looking at
> the crash, it reminded me very much of the fact that when
> I was bringing the first one out, the same thing very nearly
> happened to me in North Africa. He had exactly the same
> loading and crew as I had on the first delivery flight.
>
> About half an hour out, flying in a beautiful morning, I
> felt I wanted to go and visit the washroom, so I instructed
> my engineer to sit in my pilot's seat while I went aft. While
> I was back aft the plane suddenly did a bit of a flat turn to
> the right, then to the left, and all the luggage that we had

Top: DH 86 four-engined cabin bi-plane. Bottom: The MMA hangar at Maylands about 1935

tumbled off the hat-racks across the floor, and it was getting into a flat spin, out of control.

So I dragged myself up the cabin by the chairs, opened the cockpit door, told the engineer, 'Hop out!' and I hopped into my seat alongside the first officer. His hair was literally standing on end. I've never seen two more frightened people. The plane was directionally unstable and that was a big fault in it.

Now when I went to Longreach and inspected this crashed one, the second plane, I thought, my God, the descriptions of it! It was a beautiful fine morning, there was no bad weather, it was half an hour out of Longreach—just as when I was leaving Benghazi.

And I thought, now I wonder if he left the cockpit the same as I did to go aft. So I went into Longreach and by this time the bodies had all been taken in there and the doctor who had come out and inspected them was there, and I got hold of the doctor and I said, 'Now, Doctor, was the Captain in his seat?' 'Oh,' he said, 'look, I couldn't tell you that. I don't know them.' So I said, 'Well where did you find the bodies?' 'Well,' he said, 'there were two in the two pilots' seats and there was one right back aft near the lavatory door,' and he said 'I can tell you their injuries.' So he told me that one person had a broken thigh-bone, and the body in the front seat there had this and the one on the something there had that. So I had the grim task of going down that night to the temporary morgue in Longreach and looking at these dead bodies to see what injuries matched up with which person. And I was right!

A dreadful set-back, as Hudson Fysh described it. But QANTAS with its remaining aircraft had to fly on, and it was still a small company. As one of its younger pilots, Lennie Grey, recalled:

We only had forty-five employees to my knowledge when I joined, so they didn't have any rigid exam. Lester Brain was the chief pilot and it was he who employed me and, of course, I couldn't be employed earlier because I was under twenty-one and I couldn't have a passport before that. So to go overseas to Singapore it was necessary to be twenty-one.

But the new air route QANTAS was now flying offered exciting possibilities. For youthful pilots like Lennie Grey and their adven-

Adjusting the air vent during an early passenger flight

turous passengers, the isolation of Australia was being breached.

On our service which combined the overseas service with the Queensland internal service from Brisbane as far as Darwin, we could carry ten passengers and two pilots. Now from Darwin on, we required more fuel so passenger chairs were removed and a fuel tank was put into the cabin. That gave us the range to go to Koepang. We kept that tank in right up to Surabaya. So that we could only carry, to my memory, five passengers on that section.

Going westward to Singapore (and that was our stop, of course, we didn't crew past Singapore) Surabaya was our first night's stop in foreign country, and that was magnificent because we had a very great Dutch-built hotel there called the Oranje, and we had rooms with sitting rooms attached and magnificent dining facilities. If you had a rystafel, about forty boys came and served at the table. That was good and Singapore, of course, was excellent because we lived at the Seaview Hotel which has gone now. That was right down the waterside. We could walk to Singapore Swimming Club. We had a little golf course there, or we could go to the Royal Singapore Golf Club, and really we were very well taken care of.

Even in isolated Western Australia those international links began to have their effect. When MMA (MacRobertson Miller Airlines) acquired the air mail contract for the north-west in 1934 they began a connecting flight between Perth and lonely Daly Waters in the Northern Territory in order to link with the QANTAS–Imperial Airways service going through from Brisbane to Darwin. Alec Whitham, later to become MMA Chief Pilot after World War Two, flew that route many times.

It was a long, long flight for those days. It was twice a week Perth–Daly Waters and return. We used to fly from Perth to Carnarvon for the first day, and Carnarvon to Broome, and Broome to Victoria River Downs Station, which is in the Territory, and then Victoria River Downs Station to Daly Waters; connect with the overseas mail carried by QANTAS if it was running on time. We would wait up to three days which was the longest time we could wait because the next aeroplane would be on time to meet it if it was running that late. Daly Waters, of course, was rather a peculiar place.

Top: Keeping the heat off: at Fitzroy Crossing, Bill Bland (left), pilot Jim Branch and co-pilot Hec White rigged an awning while repairs were made to landing gear. Bottom: Refuelling in north-west Western Australia in the 1920s

It was situated more or less in the middle of Australia and had three buildings, the post office, Pearce's store where we used to actually stay and the MMA hangar. And, of course, the flies! Just literally millions of flies and in the summer time when it was quite warm, in fact it was hot!

You couldn't do very much, while you were waiting. Oh you could go out to places in the dry season and do a bit of duck-shooting, or you could drive around. But mostly it was a matter of sitting around, reading books and having perhaps some little jobs to do on the aircraft. We had to maintain our own aeroplanes in those days.

Then back to Victoria River Downs, back through Ord River Station to Noonkanbah, back to Broome, Carnarvon and so Perth. The round flight of the actual aeroplane took seven days, there and back, but the crew used to be away for ten days, where they'd have a break at Broome. Well with the crew I say the pilot. There was only one pilot.

It was worth noting that isolated Daly Waters seems to have been selected as the QANTAS–MMA rendezvous, not because it was an ideal location—Katherine would have been much better—but simply because there was an old hangar there already. Such were the exigencies of aviation in the 1930s.

In the fifteen years that had elapsed since Brearley and Hudson Fysh had first put their frail machines into the air, much had changed. Brearley had flown the longest air route in Australia and connected two capital cities and QANTAS had begun to break the barriers of distance and time between Australia and its near neighbours.

Air mail was still the major earner although passenger commitment was growing. But there were still heavy risks and costs ahead for Australia's aviation pioneers.

So Much Faith in Australia

To begin with it was always Kingsford Smith and Ulm; Kingsford Smith and Ulm in association with the *Southern Cross*.

Ellen Rogers, secretary to Charles Ulm

You know, if the dream of your life (and a mighty big dream it was, too) had suddenly come true, and for days past your life had been one long series of friendly receptions and welcomes, you would find it hard to tell the story, especially when it's a story containing so many 'I's, but now I'm going to try and tell you something of the flight of the *Southern Cross*. Right at the outset I want to stress this all-important fact. This flight was not my flight. It was a flight by Charles Ulm, Harry Lyon, Jim Warner and the chap that's talking to you.

Charles Kingsford Smith

THE SMITHY–ULM STORY

A welcoming parade in Brisbane for the crew of Southern Cross after the first flight across the Pacific in June 1928. L to R: Charles Kingsford Smith, Charles Ulm, Harry Lyon and Jim Warner.

*D*espite his *Pacific crossing* with Charles Ulm in 1928, Kingsford Smith was conscious that he was the show pony of the partnership and was at pains to stress the team work the flight had involved.

> None could have succeeded without the others. Each man had his own specialised job, and the fact that each man did do his job accounts for our presence here in Australia. It being my particular work, I was the pilot. My job was to take that splendid bus off the ground, to fly it and to bring it down again. Of course, with long hops such as flown by the *Southern Cross* there were times when I needed a spell, and then Charlie Ulm took over the joy stick.

On the flight itself, Ulm was the relief pilot.

> I sat beside Smithy in the cockpit of the *Southern Cross* with exactly the same set of controls available to me as he had. Of a total of about eighty-five hours flying, I did a little over thirty hours. My co-commander and best pal, Smithy, did the rest and naturally all of the blind flying. When Smithy was tired I relieved him at the wheel but I had plenty to do just the same.

Ulm's intensely loyal secretary Ellen Rogers also set the record straight about the teamwork.

> Coming across the Pacific created world-wide acclamation because everybody thought it was an impossible thing to do, and after they arrived in Honolulu people were all agog. Smithy could see from the newspaper reports that he was being given the credit as being the commander when in fact Ulm was joint commander with him. So Smithy sent a cable to *The Sun* and he protested at all the kudos being given to him. As a matter of fact Ulm flew almost half the number of hours it took to fly across the Pacific.

The partnership of Smithy and Ulm is one of the most remarkable episodes in the story of Australian aviation, a partnership all the more interesting for the difference in temperament, experience and background of the two men.

Charles Ulm's son John sees his father as very different from many of the first generation Australian aviators.

He wasn't, of course, a flying man. He became a very effective pilot with 2000 or 3000 hours under his belt at a later stage, but to start off with he was not a pilot. But he obviously had this business sense and the vision and the drive to make things happen. He was a soldier in the First War, but he seized upon aviation. When he was convalescing in England after being wounded in France, he seems to have made an unauthorised flight that pointed him towards aviation. But he was a remarkable guy in that (they were all mad, of course, in the early war days) he enlisted under-age; he was just under sixteen. Incidentally, Kingsford Smith tried too, but the screen picked him up and he had to wait until he was old enough. But Charles Ulm enlisted under the false name of Jackson and within a week or so of the outbreak of war. He went to Gallipoli. He was discharged after about eighteen months' service and the first discharge certificate reads: 'Charles Thomas Philippe alias Jackson Ulm, honourably discharged on being found to be a minor.' He was under-age so he came home. His father was a Frenchman born in Paris, and a real 'Tiger-Clemenceau-hate-the-Boche'-type Frenchman, and according to my mother he said to Charles, 'Well, what are you doing back here? You've still got all your limbs.' So Charles enlisted under his own name this time, and went off to France and was shot up and discharged with GSW (gun shot wound) right knee. As a matter of fact he was in pain for most of his life but he didn't show it.

Kingsford Smith also learnt to fly in World War One and in peacetime tried his hand at anything, including a spell trucking in the Gascoyne district of Western Australia. Fellow driver, Lindsay Skipworth, remembers him well.

As a matter of fact, they used to get us mixed up, we were so alike. He was a bit of a character. He used to say, 'Look, they call me "Skippy"'. I said, 'Yes, they often call me "Smithy"'. He said, 'I'm better looking than you'. He was a real hard case. He would go up to the mirror on the side of the truck and make out he was going to curl his mo. 'Oh, I'm definitely better-looking'. He was a beaut bloke, Smithy.

Ercil Williams was a typist/secretary in Norman Brearley's

Western Australian Airways office when Kingsford Smith flew with WAA in the early 1920s.

> I remember my first trip. Kingsford Smith was one of the pilots and the mechanic was Hitchcock. They worked all night on the plane and I was offered my first plane trip then, but I wouldn't get on a plane they'd been working on all night for all the tea in China! So I turned it down. My family get annoyed with me about that now because I could have said I had my first trip with Kingsford Smith, but not on your life! They'd been working on that plane all night, which makes me wonder whether a pilot nowadays could work all night on an engine like Kingsford Smith.

Gwynneth Ayers also lived in Western Australia, but down south in the wheat-belt, east of Perth. She cherishes a memory of refreshing frailty when Smithy arrived in the small country town of Tammin on 26 August, 1928.

> We'd all been very interested in his flight from England to the eastern states. And then he decided to come over to the west which we were especially thrilled about. Well, when he got here he found that he couldn't re-fuel sufficiently to get over to the east because from Perth, which is not very high above sea-level, he had to climb over the Darling Range, which goes over 1000 feet. So he decided he'd just take a minimal amount of fuel on when he left Perth, and he would re-fuel sufficiently to take him to the end of his journey, and it was at Tammin he arranged to re-fuel.
>
> He arrived there about ten o'clock in the morning and everybody for thirty miles around, or more, had all congregated there to see this wonderful machine that had flown out from England.
>
> Then the fuel company took up drums and drums of fuel. They were carted up alongside the plane and pumped into the tanks so that he would have sufficient to carry on with his journey. Of course, all the RSL were there to meet Charles Kingsford Smith and Ulm and they took him along to the hotel and gave him a very nice meal to celebrate the occasion; and afterwards they had to return to where the plane was. And I still have visions, smiling, because Kingsford Smith went to turn the propeller and he found it was a bit difficult. He was a bit unsteady on his feet after celebrating at

Top: *Charles Kingsford Smith. Bottom: The arrival of Kingsford Smith and Ulm in Southern Cross at Tammin in 1928*

this nice meal he'd been treated to and he had to get some-body else to start the plane.

On the other side of the world, in the Netherlands, KLM Captain Fiets Van Messel, who later made many flights to Australia himself, recalls meeting Smithy

> when he came here with his old *Southern Cross* to get over-hauled at the Fokker factory to prepare it for the flight over the Atlantic Ocean from east to west. That's the first time I met him, and then he was looking for a co-pilot. It was a long time ago but if I remember well, he was not much bigger than I am, small, very vivid, very active. He liked to drink like we all did; very pleasant company, very competent.

Another tribute comes from Ernest Aldis, who joined Smithy and Ulm at Australian National Airways and later became one of QANTAS's most distinguished engineers.

> He was brilliant. There's no question about it. He was what I call an airman in every sense of the word. He had a fine spirit level up in his ears which gave him the ability to blind fly in and out of clouds without the use of the more sophisti-cated gyro-operated artificial horizons we use today. He was a whimsical person. I can't remember any specific long tales but he was always the one for a good laugh.
>
> I wouldn't say he was a larrikin by any means, but business detail was the last thing he was thinking about, and getting into the seat of an aircraft and doing some flying was his prime objective. He was a bit of a will o' the wisp from the company point of view because he was out record-breaking while Ulm was behind the desk steering ANA along the way of becoming a bigger airline.

Ellen Rogers remembers Ulm doing all the work.

> Ulm was always there first thing in the morning and he worked just about every night, and Smithy would come in to the office with his pals and I don't remember in my whole association with Smithy ever having him dictate to me anything of import. He'd dictate his acknowledgments of congratulations but, as far as any serious thing, it was Ulm who dictated it and Smithy signed. But Ulm was the one who was responsible for all the organisation and the one to take notice of.

Top: *Charles Ulm.* Bottom: *In Brisbane Ulm descends from* Southern Cross *after the 1928 Pacific flight.*

It was in the practical day to day aspects of running an airline that the differences between the two men emerged most strongly. Charles Ulm's son John admired both, but believes that his father was the partner with the business head.

He set out to show that aviation was a practical commercial proposition, the sort of thing that we take for granted today. The characteristic of his life was his tremendous determination. Regardless of disasters that befell from time to time, and the setbacks (and they were many) politically, commercially and technically he never swerved in his determination.

When Charles Ulm founded Australian National Airways soon after the Pacific flight, the object of which was to get backing for a regular commercial operation, he needed an instrument-flying chief pilot, because there weren't any of those in this country and—because of his contacts in England—he approached the Chief of the Air Staff in the RAF, and Salmon was persuaded by Ulm to release one of his best instrument-flying people.

Scotty Allan, who later became a senior pilot with QANTAS and later still Assistant General Manager, also joined Ulm, and after a very brief exchange was recruited as chief pilot to Australian National Airways.

I was called to London for an interview with Charles Ulm and when I got there we sat down while he asked me questions like, 'Can you fly multi-engine aeroplanes?' Yes. 'Have you done any navigation?' Yes. When the interview was finished I departed and I didn't expect to be picked over the numerous people who applied for the job. The salary was a most enormous sum of money—£550 a year!

At the beginning Ulm and Smithy's prospects looked good. Ellen Rogers recalls that the Australian National Airways shareholders were paid an eight per cent dividend in the first six months of operations. Ulm had purchased eight-seat Avro Ten passenger planes and in nine months' operations between Sydney and Melbourne they carried 8000 paying passengers and flew over 471 000 miles.

They were flying daily each way with practically full loads, and that was in the days when people weren't used to flying, you know. They had to be made air-minded, as we called it in those days, and some of the advertising, for instance, that we had was to tell the people that you don't have that feeling of dizziness you get if, for instance, you stand on top of a building and look down. There's no such connection between the aeroplane and the earth. You don't get that. I remember the title of one of their pamphlets was: Fly— there's nothing on earth like it!

Ernest Aldis points out that ANA was, almost from its inception, one of three major players in the Australian airline business.

It was confined entirely to the east coast, operating the service from Brisbane to Sydney to Melbourne to Launceston, and later to Hobart. So there was no other airline carrying passengers or freight on the east coast of Australia. Over in the west, of course, Brearley was operating the Western Australian Airways and QANTAS were operating up in Queensland. But Australian National Airways, Kingsford Smith and Ulm's company, were the first on the east coast of Australia.

Pilot Scotty Allan recalls the ANA inauguration on 1 January, 1930.

The first flight was to Brisbane and from Brisbane to Sydney. Kingsford Smith and I flew from Sydney to Brisbane, and Charles Ulm and Paddy Shepherd flew from Brisbane to Sydney. The fare to Brisbane was six pounds thirteen shillings.

ANA's flights between Sydney and Brisbane were less hampered by bad weather than their Sydney–Melbourne run, a factor that was to prove fatal before long. Scotty Allan made the first run between the two capitals and learned early to respect the weather on that route.

On the first flight to Melbourne I had twelve passengers, but we put extra seats in the passageway to take them all. I remember well that one of the passengers was the Castle Royal Company representative and, being a very heavy man, no doubt was also somewhat nervous. We left Sydney on a day when the rain was falling heavily. Over Goulburn, which I didn't see, I was in cloud at nine thousand feet and icicles

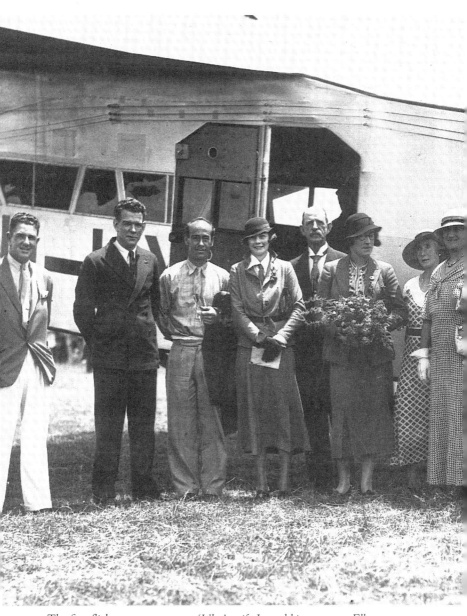

The first flight to carry women (Ulm's wife Jo and his secretary Ellen Rogers) to New Zealand. Ulm is second from left, then Scotty Allan, Ellen Rogers and Mrs Ulm holding flowers.

hanging all over the aeroplane. And I laid a course to be about thirty or forty miles west of Melbourne. I came down through the clouds. I knew the wind was against us and I would then be able to see the railway line from Adelaide to Melbourne.

I turned smartly to the left and flew along the railway line, went down, and read the name of the railway station, Sunshine. I therefore followed the railway line till I came opposite where Essendon was. I turned to the left and landed at Essendon and taxied up to the hangars. It was still raining. The visibility was still bad and I stopped.

ANA engineer Ernest Aldis was well aware of the dangers of this route in wintry conditions.

The aircraft had no radio at all and there was a high degree of blind flying when in cloud and out of cloud. There was no meteorological service in those days. If you take a direct line from Sydney to Melbourne you go right over the Australian Alps, very inhospitable country, and it was a flight which required a compass course setting, using dead reckoning when you're in and out of clouds, to estimate your drift and your track position. The aircraft frequently ran into icing conditions, and the aircraft had no anti-icing or ice removal equipment installed and so, compared to the Brisbane run, Australian National Airways always put the best airmen on this run.

Such weather was not uncommon on the Sydney–Melbourne run, and it was in just such bad conditions that another ANA aircraft, the *Southern Cloud*, flown by Captain TW Shortridge, left Sydney on 21 March, 1931. Engineer Ernest Aldis was on the tarmac.

In fact, I cranked the engines up before it left, and little did I know, of course, that the aircraft would do anything other than fly from Sydney to Melbourne like all the other aircraft had been doing quite regularly.

At the same time Scotty Allan was bringing his Avro Ten up from Melbourne.

That was an extremely bad day. Ships were hoved to outside the harbour. I came up from Melbourne that day, of course, and I took off at Melbourne, raining, low cloud. By the time I got to six or seven thousand feet, ice all over the aeroplane, in cloud, and I never saw the ground.

The wind was blowing at the height I was sited, a hundred miles an hour and it must have been sixty or seventy miles an hour near the ground. So that Shortridge going south heading into the wind in a aeroplane that only did ninety miles an hour, wouldn't make much progress in flight. If he decided to go out over the Riverina Plains, the aeroplane would just go sideways.

When he himself landed at Sydney, Scotty Allan got into his car and set off for his digs at Kings Cross. Charles Ulm was waiting to tell him that the *Southern Cloud* hadn't arrived in Melbourne, and it was well beyond the time that its fuel would have run out. For Aldis the first reaction was shock.

We were all stunned to think that the aircraft we sent off in the morning had disappeared and we expected somebody on the ground would have told us that the aircraft had landed there for fuel or diversionary reasons, but we just couldn't believe that the aircraft had just disappeared into smoke.

The first action was to get the remaining aircraft we had at Mascot heading down over the route to see whether we could spot them, and one of the little tricks we did on these Avro Tens was to cut a hole in the floor (which was a plywood structure) big enough to lay on the floor, because there were many hours of flying in the search area, and we looked through the hole in the floor throughout the trees as we went through.

But it was to prove a fruitless search. The *Southern Cloud* would not be found for a quarter of a century.* Charles Ulm, Ernest Aldis, Scotty Allan and other ANA staff worked around the clock, flying backwards and forwards over the Melbourne–Sydney route. But in the end Ulm had to accept that the *Southern Cloud* had disappeared.

---

*The wreckage of the *Southern Cloud* was discovered in the Snowy Mountains in October 1958 by a photographer, Tom Sonter, who was heading for the nearby Tooma River gorge. The engines had buried themselves steeply in the ground and there was evidence that the aircraft had burnt fiercely on impact. A watch found in the wreckage had stopped at 1.15 pm on 21 March, 1931.

Ernest Aldis has speculated many times since as to what happened to Shortridge and the aircraft.

One of the big hazards of flying over that area, particularly large-wing aircraft like the *Southern Cloud*, was the build-up of ice. If they struck icing conditions, which I presume that they did, the aeroplane would have been loaded up with extra ice weight.

Also the ice has a habit of changing the aerofoil shape of the wing to reduce lift. So the aeroplane would stall and virtually fall out of the sky. He had plenty of fuel so we can only put it down to the fact that weather conditions produced icing and it upset the aerodynamics of the aircraft and caused it to stall and nosedive.

The loss of the *Southern Cloud* with its crew and passengers was possibly the factor which decided the fate of Australian National Airways. Scotty Allan believes the airline went broke after the tragedy, spending scarce capital in the search for the lost aircraft. At all events ANA suspended its services on 26 June, 1931, after one and a half years of operation.

As Ellen Rogers points out, they were operating without any government subsidy whatsoever

because Ulm said he knew that the company could operate without a government subsidy. He had so much faith in how the public would react to regular services.

But it was not just the loss of the *Southern Cloud* which caused the airline to collapse. The world-wide Depression was a major factor. The demand for travel and especially business travel had dropped sharply.

Despite the loss of his airline Ulm was undeterred. With his own money he bought and rebuilt *Faith in Australia*, originally an ANA Avro Ten, the *Southern Moon*. He wanted to show that Australia could build aircraft capable of handling the long distances required to secure the England–Australia air mail contract, now being vigorously pursued, at the Australian end by QANTAS. With former ANA colleague Scotty Allan and experienced pilot PG Taylor he planned to fly around the world on a promotional flight. *Faith in Australia* reached England

Top: *The ill-fated* Southern Cloud *at Mascot. Bottom: In this photograph Charles Ulm and Travis Shortridge, later to lose his life in the* Southern Cloud, *are standing together on the left.*

after seventeen days and then prepared for a major leg of the trip, the Atlantic crossing from Ireland. But, here, as John Ulm describes it, their luck appeared to have run out.

The aeroplane was fully loaded with fuel at Portmarnock beach just outside Dublin for the Atlantic crossing, when probably the shifting sands under the wheels led to one of the undercarriage legs collapsing and the aircraft was wrecked.

Now that was the end of Charles Ulm's personal resources, and he and Scotty Allan and Taylor [Bill Taylor] went back to the hotel. Charles Ulm said to them, 'Well look, you chaps, you go on home. I don't know how I'm getting it out of this, but I will'. As they were talking a telegram boy knocked on the door and he handed a telegram to Charles Ulm saying: 'I will be happy to bear the cost of putting your aircraft back into the air. Signed Wakefield of Hythe'. Lord Wakefield was head of the Wakefield Oil Company.

Despite the reprieve, Ulm was still unsuccessful in winning the Australia–England air mail contract. That went to Imperial Airways and QANTAS in 1934. But Ulm persevered. In the 1930s it was still hard to sell aviation to anybody, especially to politicians, but later, when Ulm had established yet another venture, Great Pacific Airways, Prime Minister Joe Lyons was sufficiently convinced of the value of Ulm's plans to link the United States and Australia that he sought and obtained unanimous Cabinet approval to guarantee Charles Ulm's bank overdraft for £8000 to finance the development flight.

It was to be Ulm's last journey and, had his earlier hopes been fulfilled, a flight he should not have had to undertake.

Ulm set out in his aircraft, *Stella Australis*, from the United States to fly to Australia, and was never seen again, lost on 4 December, 1934, somewhere between the west coast of America and Honolulu. Less than a year later Smithy was also to lose his life over water.

Aviation historian John Gunn estimates the price both men paid for their achievements.

Smithy and Ulm depended for their influence on their public reputation, gained by doing spectacular and heroic things. They had constantly to reinforce their reputation as more and more people joined in and made these great long-distance flights. For example, both Ulm and Kingsford Smith made separate high speed flights from England to Australia within weeks of the issue of the conditions for the air mail contract. Smithy flew a Percival Gull called *Miss Southern Cross*. He flew it to Australia in seven days, and to give you an idea of the public perception, there was a crowd of about 100 000 people greeted him when he arrived in Melbourne. That was in October, 1934. Ulm left three days before Kingsford Smith arrived, in his aeroplane *Faith in Australia*, and he bettered the time of the Gull. He did the trip in six days seventeen hours.

And following that through to its bitter end, Kingsford Smith left London in November 1935, in the *Lady Southern Cross*, to attempt to break the record again, and on a night crossing of the Bay of Bengal, Smithy vanished.

In Gunn's estimation, there are only a limited number of top people in any field at any one time and Ulm was one of those.

So we lost somebody who was not only a great pilot and had all those particular qualities in the air like Smithy, but he had a structured intellect. He was a man who could express himself well in words. He could quantify things in figures and set them down and give proportion to what otherwise might be just phrases or aspirations. In other words, he was a born leader and manager, I think, and I'm quite certain he would have had his own airline. It would have flourished and he would have somehow changed the shape of things.

QANTAS founder Hudson Fysh knew both men well. Ulm had always been a potential rival but he had unstinted praise for Ulm's partner.

Kingsford Smith's record as an airman is unique. There's no other aviator in the whole world with his record in regard to the conquering of the great oceans, flying over the great oceans of the world.

And, from Ellen Rogers,

I had the greatest respect for Ulm. He was a very considerate boss and he was absolutely absorbed in his work. He was a genius, really.

Top: Faith in Australia *in Manchester, UK, in 1933 with Captain Scotty Allan, Commander Charles Ulm, Navigator PG Taylor and Wireless Officer Edwards. Bottom: Scotty Allan flying the* Southern Sun *over Sydney in December 1931 at the start of the first air mail flight to England and return*

But perhaps their own words, jointly delivered after their joint Pacific flight of 1928, are their mutual epitaph. From Charles Kingsford Smith:

> Our trip was like all other big trips. We had bad times and we had good times.

And Charles Ulm:

> There were times when we were down to zero, but we managed to stick with the job. We knew we could make it successfully if we could only get the support necessary and the right equipment. We knew that unless every link in the chain was strong and rightly placed we were fore-doomed to failure.
>
> We're home now. The job is done. It was teamwork and loyalty to the job in hand that pulled us through. I've learned this all-important lesson which must be of value to other fliers. It is this: when you've got a job to do, look after the little things.

That Bloke's the Flying Christ

At the first race there was a stockman from a nearby cattle station who was thrown right at the finishing line. And two horses fell on him and his head was split open and a leg broken.

I put up my pedal radio and called the Flying Doctor, 200-odd miles away in Cloncurry, and we waited for the drone of this aircraft in the distance, not knowing what to do with this fellow. Well, the plane came and landed and the doctor, he just rushed straight to the patient with his bag and he was giving morphia and making splints. He put this fellow in a stretcher and he was carried to the aircraft and he was away again. As he took off with his pilot, he flew over the eight furlong track and I was standing with two old bushmen, two old characters of the outback. One was smoking a stub of a pipe which had broken off, and he had this in his mouth, and as the plane passed over us the shadow came of the cross of the aircraft, and this old bloke he took the pipe out of his mouth, and he pointed it up to the aircraft and said, 'That bloke's the flying Christ'.

The Reverend Fred McKay, former superintendent of the Australian Inland Mission on a field trip in Central Australia

EARLY DAYS WITH THE FLYING DOCTOR

As a young man, Fred McKay had no thoughts of serving the people of the inland, but almost before he realised what was happening he was roped in by the AIM founder John Flynn.

When I finished my university work in Brisbane, 1932, twenty-five years of age, they sent me to Southport, beautiful place, seaside. I'd done a fair bit of swimming at the university and I loved the surf club at Southport and I felt that this was paradise from my point of view. And after I'd been there for two years and become a real part of the wider community, too (because Surfers Paradise was quite a different place in those times), John Flynn, this intriguing statesman of the Church, came looking for a recruit to go out to Cloncurry and Birdsville and thereabouts.

I had no idea what his purpose was. He turned up in the train and I was batching at the back of a church, and on the Sunday he took my services. We travelled round and did a round-up of the parish, and the country places, too. He was talking and I was listening and I wasn't greatly interested. I had plans of going to Edinburgh to do some post-graduate study and my dreams were set in other directions.

On the Monday morning, which was my regular habit, I went to the beach and he came with me and he was dressed in his traditional way, a waistcoat and hat and a watch chain across his waist and he sat on the sand and I surfed. And I came back and sat beside him and he talked. John Flynn was a great talker, and he was a great listener. I mean he would listen to people and delve into their interests. He went on to talk about Birdsville where he'd started a nursing home in 1924 in an old pub, and he said, 'We have to build out there. We can't carry on like this because it's not fair to the nursing sisters, and,' he said, 'we have to penetrate up to Cape York.' He said, 'We want a young fellow to come out and mooch around the Gulf and the Cooper Basin country,' and be sort of his agent with the Flying Doctor Service and the pedal radio and the nursing home work, and also be a normal kind of itinerant padre amongst the people.

I was playing in the sand. I loved the sand at Southport— beautiful sand. And he took a handful of it and he let it run through his long fingers and then he said to me, 'Fred, the sand out at Birdsville is a lot lovelier than this.'

I don't know what happened. We walked up from the beach to the dressing shed, a narrow sort of path up near the shed, and he had a loping stride, stooped shoulders a bit, and he just loped on. I can still see him doing it. And I walked behind him and I put my footsteps in the marks he was making in the sand, and I think from that moment they've been there ever since.

The flying doctor story is inseparable from the development of outback civil aviation. QANTAS founder Hudson Fysh was both a Flynn convert and also a business partner. In fact, a flying doctor service of sorts had begun in the early 1920s, along with Hudson Fysh's own efforts to get QANTAS off the ground in north west Queensland. In one form or another flying doctor services sprang up in many isolated parts of Australia. It was Flynn's genius to unite them with the work of the Australian Inland Mission.

Aviation writer Ted Wixted recalls one of the earliest visionaries.

There was a chap in Brisbane, Hugh Davis, who was the secretary of what is now the Royal Queensland Aero Club. In 1922, before he knew anybody else had the same idea in their heads, he gave a public address in Brisbane on the need for flying doctor services in the outback.

The man who took up that challenge was Dr Michod. Longreach local historian Angela Moffatt traces his contribution to the inception of QANTAS back in 1921.

It was something that was begun almost immediately. The need was there and cars were already proving quite a boon. The same problems were being experienced as they had with horse-drawn vehicles. When it rained, once it rained, you just didn't go anywhere even if you had motorised transport. You just couldn't move. Whereas a plane in most cases could get through.

Long-time Longreach resident Nancy Button was one of his early station patients.

In my early life there were lots of eye infections out here with the dust and the grit and everything, and you got sandy blight. Your eyes would be very bunged up. Anyway, then it would turn into trachoma and Dr Michod, he was most interested in this, and I can remember going every morning and getting my eyes painted. They'd turn them back and they would paint them with blue stone. He did a lot of research on trachoma. Dr Michod was a real institution here.

Michod was one of the early directors of QANTAS for a number of years and saw the advent of flight in central Queensland not just as a passenger service but also an ambulance service. If planes could bring the sick or injured from an outlying property to a base hospital, they could and did save lives. Hudson Fysh shared many of those mercy dashes.

I remember in 1924 I flew into Longreach a lady from Corona Station who was beset by the floods and going to have an infant at any moment. There was a great hubbub and commotion with Dr Michod at Longreach. Would I go out and get her? So I went out and landed on a little ridge alongside the station and we flew this lady back to safety. Of course I was very frightened that she might have the baby coming along but we got in all right and to my knowledge I think this was the first maternity case which was flown into hospital in Australia. That was in 1924.

Hudson Fysh had agreed from the outset of QANTAS to supply both aircraft and pilot. He was adamant that the doctor should not fly his own plane. The conditions of landing on poor airstrips and the possibility that the doctor might have to do medical work in the air, including delivering babies, made the service of a professional pilot essential. His view was endorsed by pioneer flying doctor Alan Vickers.

As a policy in the Service, we are very determined that the team shall be a two-man team and that the flying shall be done by a commercial pilot with not less than a thousand hours commercial line experience. The type of country is not the country an amateur pilot should fly and we feel that a commercial pilot of that experience is a much safer proposition.

Years later, in a 1946 broadcast, Dr Alan Vickers recalled

being the only flying doctor in Australia. But nowadays I'm

glad to say I have six colleagues because we have at last got our seven bases where we intended to put them, and regard our Australian Emergency Service as fairly complete.

Nancy Button remembers Alan Vickers in the 1920s as a man to whom everybody was important.

It didn't matter how humble their beginnings or whatnot. You'd have an old chap on a station. He might only be a boundary rider or anything, but it was just as important for him to look after the boundary rider as it was the owner of the station.

I remember Dr Vickers was telling us about this chap who had a very bad toothache. They were in communication with each other by telephone, and Alan Vickers was telling them how to go about pulling this tooth out. They couldn't get out there because of the boggy roads. So this chap said, 'Oh well, I'll have to do it. I'll have to pull it out.' So he said, 'What do I do?' And he said, 'Well, you get your multi-grips or something like this, and you pull it'. He said, 'I think the best thing to do would be to give your patient a good rum before you start'. And so this chap said, 'oh that might be a good idea'. He said, 'I think I might have one too. I'm feeling as nervous as the man himself.' So anyway they both had a rum and they pulled and nothing happened, and Dr Alan Vickers said, 'Well, try and do it this way and have another go'. So he said, 'I think we'd better have another rum,' and he said, 'Well that might be a good idea'. I don't know if the tooth ever got out, but the patient wasn't feeling any pain and the chap that was working on the tooth, he wasn't worrying about anything at all!

Pilot skill, medical knowledge and a sense of community co-operation kept the whole venture alive in the early days. Landing grounds were often patches of uneven ground which, with any luck, didn't put too many trees or rocks in the pilot's landing path. The Shell Company willingly assisted with fuel dumps at unlikely spots.

The aircraft could land on rough places and on roads, but they could not fly at night or under flood conditions where airstrips were boggy. But this arrangement between QANTAS and the AIM lasted for a vital formative eleven years.

Pilot Paul McGinness with the family of a little girl helped by the Flying Doctor at Longreach in the early 1920s.

The Reverend Fred McKay saw both altruism and economic common sense in the arrangement.

Hudson Fysh, whom I knew well and had a lot to do with in later years, he was a hard-headed businessman as well as a professional aviator from out of the War, with a great record; and he and McGinness, of course, set up this thing in Longreach and his deal with Flynn was a contract. He wasn't going to lose money. He said, 'I'll do this for two shillings a mile'. And the AIM actually helped to put Hudson Fysh on his feet because he had the mail contract, but this wasn't a big money winning project. The Flying Doctor Service really was a guaranteed income and he was able to combine it with his mail contract in Cloncurry, too, because the same plane did the mail contract. Now John Flynn had a great motto: 'A man is his friends'; and he had made a real bond of friendship with Hudson Fysh of QANTAS.

Hudson Fysh and Flynn planned the beginning of the Flying Doctor service over many talks in the old Metropole Hotel, Brisbane. An innovator himself, even Hudson Fysh was sometimes left behind when it came to applying technology. He wasn't sure how the service could be started.

We didn't have an aeroplane then that you could put a stretcher in. Flynn was difficult because he had a one-track mind on the air ambulance. He wouldn't take a no. If you said, 'Well it's not economical,' he'd say, 'Well, we've got to have the plane,' and we really had to wait until an aeroplane came along in 1924.

Deaconess Frances McKechnie, for many years Executive Officer to the Australian Inland Mission, describes the sequel.

Hudson Fysh told some of us, just before he died, that Flynn said to him he wanted an aeroplane, not with two cockpits, but with a cabin; and he said, 'Oh John, don't be ridiculous. You've got marvellous ideas but you know nothing about aerodynamics.'

'Well,' he said, 'What happens if a man's injured out in the middle of nowhere? How do we get him to a hospital? He will die jolted over the roads in a truck or something.' So Hudson Fysh went to de Havilland. They laughed, and he said, 'Look, I'm just telling you there's a parson back in Australia who has no idea of taking no for an answer, and if you don't come up with it he'll go to someone else.'

In 1924 de Havilland came up with the aircraft Flynn and Fysh had ordered.

It was the DH 50, with a stretcher, and John Flynn said, 'That's it! That's it! We can put the patient in that stretcher and we'll have the doctor sitting alongside him,' and that's how we got together and QANTAS did the first contract with the AIM in 1928 in Cloncurry.

The British-built de Havilland 50 was a water-cooled single engine bi-plane. The pilot sat outside and used a tube to communicate to the doctor, frequently shouting above the noise of the engine. But the great advantage of the DH 50 was its large side door through which patients could be loaded on a stretcher. In addition the aircraft could carry two or three sitting patients, sometimes four, if small folk.

These planes were not only ideal aircraft in their time for medical transport work in bush areas, but there was another bonus: Hudson Fysh had secured a contract from de Havilland to put them together in Longreach.

The Australian Inland Mission had begun with committed doctors and had acquired the use of aircraft and skilled pilots. Frances McKechnie believes that Flynn was born at the right time.

It was just as motors were coming in. It was just as the first aeroplanes were being used; wireless was coming into use and these things he could see transforming the outback. I don't know that anyone else saw their application for the outback. Then there was the pedal wireless. There were four things he said about it: it must be simple enough for a child to operate, light enough for a woman to lift, powerful enough to receive and transmit a message five hundred miles, and cheap enough for the poorest to buy.

The way Fred McKay saw it, Flynn's talent also lay in finding and using the right people. Adelaide engineer and inventor, Alf Traeger, was the genius behind the pedal radio.

After experimenting for years, two years at least (1927 and 1928), in 1929 he was ready to install these midget miracle pedal transceivers which he did up to Leichhardt, and once these sets had been installed, bush people began to realise, this it it! We've got communication. The old dumb bush is starting to talk.

The invention was tested almost at once.

By the time Traeger had got through with the pedal wireless, de Havilland had put the first cabin in an aeroplane and QANTAS had it operating. The plane was standing by in Cloncurry when they were setting up the pedal wireless. A horseman galloped in—his mate had been injured—and they flew to what's now Mount Isa. He heard there was a doctor there—it was Dr George Simpson. They picked up this man, flew him to Cairns and saved his life.

In 1929 the first base station was installed at Cloncurry. The first sets, although they were pedal-operated, could only communicate in Morse. The furthest out station at the time was Birdsville, 700 kilometres from Cloncurry.

The Cloncurry base itself was well planned, well supported and almost immediately provided a sense of security for inland people. But as Alf Traeger himself recalls, early communications were difficult. The signal was all right,

but it was Morse from the base station. The operator had to read the Morse from the pedal sets and some of that Morse wasn't so good because they weren't Morse operators.

But Traeger would soon solve that problem, too. As Frances McKechnie pointed out, the interesting thing about Traeger's invention at this point was that he had everything right, except that

you needed two hands to operate the machine so how did you generate power? And of course it was while he was going home from work one night on his bike that he suddenly thought, pedals! That's the answer.

Flynn put the whole of the money he'd collected, which was for his salary, stipend, into Traeger's work. I guess you wouldn't say he was a gambler but I think he was a tremendous gambler. He put everything into that. At another time when the Flying Doctor base was operating out of Cloncurry and the call came from Mornington Island, the doctor asked and he said, 'Go! If you use all the money we've got, go there. They need you. That's all that matters.' And he was prepared to risk everything they had for the things that mattered, and in the end it worked out all right.

Fred McKay remembers an incident where the gamble paid off. The year was 1934 and the incident made aviation history.

Dr Jock Russell, just a fortnight before, had travelled in the old DH 50, 600 miles from Cloncurry down to Innamincka, the Burke and Wills country, re-fuelling at Boolya, Birdsville, Betoota; the longest trip actually that the Flying Doctor had done up to that point of time. On the aircraft on that particular flight, the doctor took with him Maurie Anderson who was our head wireless operator in Cloncurry. Every time he flew he had a radio set and he was trying to contact base, you know, back in the mother station, and on this particular flight for the first time in medical history, the doctor was able to give advice and guidance to the two nursing sisters in Innamincka, being relayed from Cloncurry over to Innamincka.

The doctor arrived later that evening and the patient recovered. The incident was good publicity and Flynn was a man who knew the value of goodwill. Fred McKay also saw him as a builder of networks.

On the radio side, the work that they did year after year, anyone would have given up, and also getting resources for the Flying Doctor Service. He got Hugh Victor McKay of Sunshine. He developed a great friendship there, and he got a man called Barber, Andrew Barber, to go round collecting resources to start off the Flying Doctor Service.

Frances McKechnie knew Andrew Barber's son.

Yes, I love the story that Judge Barber tells when Flynn visited his late father, the Reverend Andrew Barber. The Judge was a little boy and he was riding his bike which was squeaking, on the front path, and Flynn hadn't even met his father, but he stopped to fix the squeak in the bike before he went in to meet the father. And as Judge Barber said, 'I was Flynn's man from then on'.

Flynn himself was keenly aware of the social as well as the medical benefits of the pedal radio. In one of his radio broadcasts he addressed a potential source of complaint about the service itself.

> I think you who've been following our progress, and interruptions to progress, out there, would be glad to hear something of the atmosphere of that lonely land as it has been modified by the wireless transmission side of the Flying Doctor Service of Australia. Now you may have heard people discussing this Service in the past, rather bitterly criticising it because the neighbours could hear your business. There may be times, especially if you're interested in things that other people want to know about particularly, there may be times when you would rather they didn't know your business. But other people (and you may be amongst them) don't realise that in most things the knowledge your neighbours have of what you were doing and what you were talking about is a safeguard to you.

Fred McKay confessed himself quite unprepared for the way the radio had already transformed the back country.

> Now at Birdsville, I suppose, eight miles from the border fence out there, the girls, two nursing sisters, were working in a hotel which they'd set up as a very nice clinic and wards and hospital equipment and a pedal radio, and there I heard for the first time the 'galah session'. These sisters talking to the womenfolk all round the countryside the next morning (early in the morning) and the things they talked about. It amazed me because it was social history plus, because you began to realise what made life tick. I used to listen in for pure science's sake, just to find out what women talked about. They would talk about the making of their bread. How has your bread turned out this week. And what about that catalogue from David Jones? Have you seen that new frock on page 26? And what about the kids, how are the correspondence lessons? Can you get that sum out, you know, that was in Form Two yesterday or the day before? This actually enabled the mothers who were virtually the educationists in the homes—and, of course, the homemakers in every home—they were able to share problems and it was exciting, yes.

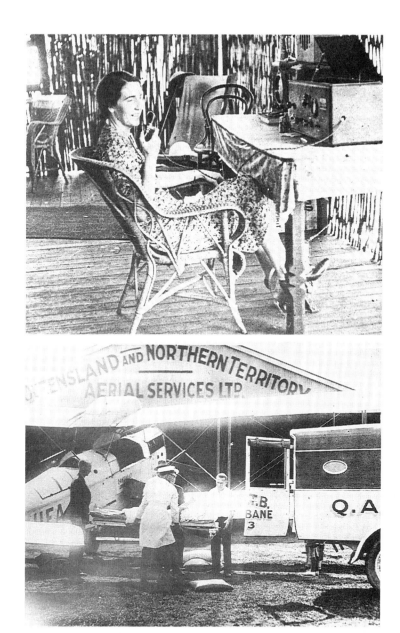

Top: Mrs Irene Fuller of Newry Station in the Northern Territory using pedal wireless to contact the Flying Doctor Service. Bottom: The Flying Doctor at work

Frances McKechnie felt that the shortwave radio gained quick acceptance but earned its best brownie points when the Melbourne Cup came over the shortwave for the first time.

One or two families realised the value and you only needed an emergency and there was no resistance to it. I think at first they just wondered what it was, but just as it began, a nurse was setting up the base at Cloncurry when a man came in searching for a doctor and people realised that Flynn's ideas were very sound. Then when they heard the Melbourne Cup broadcast, that was wonderful.

A broadcast that, as Alf Traeger recalls, nearly went off the air.

The race had almost finished and the reception was suddenly interrupted by a dog fight that broke the aerial wire and, of course, Mr Flynn was right on the spot. He saw what had happened and he quickly connected it and the situation was saved by the placings coming through loud and clear.

For Fred McKay there was one day that seemed to capture the whole spirit of the Flying Doctor ideal.

I went to Gregory Downs, to a big race meeting, and Flynn said, 'Never fail to go to race meetings because that's where you meet all the people'. And this was in my early experience and I was quite a rookie at the work.

I turned up and in the first race there was a stockman from a nearby cattle station who was thrown right at the finishing line. Two horses fell on him and his head was split open, a leg broken, and we covered him with leaves and so forth.

I put up my pedal radio and called the Flying Doctor, 200-odd miles away in Cloncurry, and we waited for the drone of this aircraft in the distance, not knowing what to do with this fellow except we were giving him water and so forth but nothing else.

Well the plane came and landed, and the doctor, I've never seen such an unpretentious little drama in my life because he didn't even speak to me. He just rushed straight to the patient with his bag and he was giving morphia and making splints. He put this fellow in a stretcher and he was carried to the aircraft and he was away again. There were no cameras and no reporters, no ABC there to see what was going on.

And as he took off with his pilot he flew over the eight furlong track and I was standing with two old bushmen, two old characters of the outback. One was smoking a stub of a pipe which had broken off and he had this in his mouth, and as the plane passed over us (it was just after mid-day) the shadow came almost at our feet, of the cross, you see, of the aircraft. And this old bloke, he took the pipe out of his mouth and he pointed it up to the aircraft in the sky and he said, 'That bloke's the flying Christ'. Now I never heard the doctor being described this way and the doctor would be a bit embarrassed to be described that way, but that was Flynn in action. You know, putting something there that even to an old rugged bushman would have a meaning, and they'd say, 'oh gee, there's more behind that than just a bloke putting a fellow on a stretcher'.

The mythology of the Flying Doctor has been built on images like that, the injured stockman, the pregnant young mother, archetypal outback characters with their share of romantic associations, all wholesome outback characters and all white.

John Flynn has been criticised in recent years by some who felt that he only dealt with European Australians, a view vigorously refuted by Deaconess Frances McKechnie.

I remember one Aborigine coming to me and saying, 'The day that John Flynn and Traeger set up the wireless on Mornington Island, that day we stopped dying and began to live'.

Hold the aircraft steady, we're putting.

Hold the Aircraft Steady We're Putting

Taking off, they had that lovely rising from the water with this lovely stream flowing back.

Gladys Ritchie who joined QANTAS in 1938.

I loved them. They were like great big dolphins, you know, they were friendly. You could almost speak to them. I really got on quite good terms with the flying-boats. I thought it was just unbelievable that something so large could be kept up in the air with four engines, and not fall down.

Wendy Myles, daughter of QANTAS founder Sir Hudson Fysh, recalling childhood at Rose Bay.

The view was very, very good for the passengers because firstly you weren't away high up; secondly it was a high wing aeroplane which meant you could stand on the 'sports deck' as we used to call it, and look out of chest high windows, which was very pleasant. A lot of people used that deck. They also used that area for golf putting and occasionally we got a message on the flight deck, 'Please hold the aircraft steady, we're putting'.

Veteran QANTAS pilot, Lennie Grey

FLYING-BOATS AT ROSE BAY 1938–1939

*T*he flying-boat era was an all too brief part of Australian aircraft history, starting in the 1930s but scarcely surviving World War Two. For QANTAS, flying-boats came with the partnership with Britain's Imperial Airways, a partnership that created a new name, QANTAS Empire Airways and involved another change of base from Archerfield, Brisbane to Rose Bay, facing Sydney's immense harbour.

An outback airline which had only just begun to operate multi-engined aircraft was now taking on state of the art aeroplanes, the magnificent new flying-boats. That meant new routes, the long hop over water to Singapore, new training, flying-boats required additional pilot skills.

The move from Brisbane to Sydney had to be completed by August 1938, when the full flying-boat service to London via Singapore would begin. Lennie Grey was to miss the opening ceremony on 4 August.

> I was still shuttling back and forth to Singapore in 86s right up to the end, I think, and we were also shifting most of our worldly goods from Brisbane to Sydney. We were using 86s for that and we were doing training flights in the flying-boats and filling them up with office equipment and things like that. And that took some months. We had two flying-boats out in Brisbane and one of Imperial Airways' captains was there, Captain Lynchbloss. He and a couple of our own people went through pretty quickly and we did the training there around Brisbane River and Moreton Bay. That gave us all we needed plus a bit of sailing boat practice and things like that.

Training, or rather re-training, was intensive. Many of Lennie Grey's contemporaries had been flying bi-plane DH 86s and even older aeroplanes. Apart from a few who had gone to the Hythe flying-boat base at Southampton in England and the training in Brisbane, there was little watercraft experience at QANTAS.

But Keith Calwell had already flown amphibian aircraft when he joined QANTAS in 1938

Flying in calm water is quite a different thing from flying in

operational conditions in various ports and rough seas, and
that sort of thing. So I had the opportunity on the flight up
the coast to really learn what it was all about and learn what
you could do with a flying-boat.

Engineer George Roberts had joined QANTAS in Brisbane in
1936. He was involved in practically all test flights for the first
three and a half years.

> Every test flight was normally around about an hour. It took
> us up the coast or down the coast or wherever it might be.
> Certain adjustments had to be made in the air because we
> did not have the means of doing it on the land. One of them
> was crawling out in the wing behind the motors, in the wing
> in flight. If you think in terms of the wing section as the
> centre section. It was about five feet deep. So there was a
> little door on either side and you crawled through that door
> and then out behind the engines. It was extremely noisy.

Noisy too for those on the ground, or rather those near the
water once the planes began to operate at Rose Bay. And in
1938 the bay was very much a residential area, much as it is
now. Residents were worried by the thought of noisy aircraft
coming right into the harbour and landing in front of their
waterside homes. A resident resistance group formed to protest
the new development. Gladys Ritchie was soon, by her own
admission, a biased witness.

> As far as I can recall there didn't seem to be much noise at
> all, but of course I was used to them. There was a lot of noise
> down at the flying-boat base, what with engines being run up
> and us trying to answer the telephones while the engines
> were running, so probably when I got on board it almost
> seemed quiet. It did alter the appearance of that particular
> part of the bay, but that was only a small percentage.

Hudson Fysh's son, John, then in his teens, was well aware of
reaction to the new aircraft.

> The people there were very frightened that it was going to
> disturb them and be noisy and I suppose to a little extent it
> did, but those flying-boats were very quiet and, in fact, when
> the first one arrived, the demonstrators such as they were in

those days, the people who were there to see how much it was going to disturb them, didn't realise the aircraft had already landed and was on the water.

The Fysh Sydney home, a flat at the top of Point Piper, over-looked the harbour and Rose Bay. For John Fysh's sister, Wendy, it was a bird's-eye view.

My father used to take me down quite often in the mornings. He was trying desperately to make a swimmer out of me so we'd go into the Rose Bay basin, see that all was well before the planes took off and then, if I was lucky I'd get a ride in one of the Civil Aviation boats all over Rose Bay, which was just brilliant. It was so beautiful, and those misty early mornings in Rose Bay … it was a lovely time.

For passengers, as well as onlookers, the flying-boats offered a new dimension to flying. They were bigger than any aircraft yet seen in Australia, but graceful in outline and reliable in perfor-mance. Above all they were comfortable. Aviation writer John Gunn describes their impact.

The flying-boats were so extraordinarily spacious you could get up from your chair and you could stroll to what was called the 'smoking cabin'. You could stop on a promenade deck. It had windows at eye level and you could look out at the clouds. You could even play quoits or you could play clock golf there.

And they were beginning to serve very sumptuous kinds of meals: grapefruits, cereals, eggs and bacon for breakfast, and roast mutton with peas and potatoes, peach melba. It was pretty good service. I don't think we'll ever see the like of it again.

That view was endorsed by skipper Keith Calwell.

Oh they were wonderful! The flying-boat had a 108-foot wing span and I think it was 97 foot length of the hull. The inside of the passenger compartments, like the promenade compartment, was about 10 foot 6 from floor to ceiling, and quite a wide space, say about 5 or 6 foot wide, by 12 to 14 foot long for anybody to come and walk around in.

Lennie Grey was delighted with the extra space.

The feeling we got was that you had a very much larger aircraft than what we had previously, a spacious aircraft. Not so many more passengers but a very big spacious aircraft—all metal You felt it was totally different to the wood and rag days of the earlier aircraft.

The excitement of the new aircraft was infectious. Office staff like newly recruited Gladys Ritchie found them too distracting at times.

Mr Arthur Baird, who was the works manager there and our boss, saw us looking out the window watching one of the boats taking off—this elongated tin shed that we worked in was quite close to the foreshore. If we didn't have our heads down working, we could get up and look out the window, and this other lass delayed a little longer than myself and he took her by the waist and introduced her to her typist's seat and told her to get on with what she was supposed to be doing, in a very kind gentle way.

Gladys Ritchie had just completed a business college course as QANTAS set up its Sydney base. Within six months she was part of the big move. Much of the work at this stage was technical, inter-base memos, long lists of spare parts that had to be typed out. But Gladys was soon to have the pleasure of going up on early test flights.

Of course it was just a dream come true to be able to go out on a flying-boat. We went down, of course, to the waterfront and walked along a small gangway on to a pontoon, then boarded a small ferry boat, quite a small one because, after all, the flying-boats only held about fifteen passengers, and then we were taken out to the aircraft which was moored on the water, and then we just stepped in to the flying-boat itself, and they were lovely.

Behind the glamour of the new technology there had been a lot of hard work and, from the British side of the partnership, an element of intrigue. Britain had already made imperial assumptions about carriage of the England–Australia mail service which had begun in 1934. Now Imperial Airways was to make similar assumptions about the aircraft for the route. Just before scheduled flights with land-based aircraft were due to start, Imperial Airways came up with a new idea. They would order, off the drawing board, a great fleet of big flying-boats.

The aim was an air mail service which would link all parts of the Empire with no surcharge for the carriage by air. But, as John Gunn sees it, they took Australian consent for granted.

They simply didn't tell the Australian government, and they presumed we would go along with it. The Australian government saw it as a plot for British control over the whole route and they, in fact, fought it. QANTAS, under McMaster and Fysh, could see the merit in the flying-boat and in the idea, and went along with it.

Eventually it was resolved that QANTAS would operate the Singapore to Sydney section of the Empire route, with Short C class flying-boats.

The new routes and the new type of aircraft brought their share of problems, problems that had to be solved by engineers like George Roberts.

When we arrived here there was no hangar and maintenance had to be done on the water. If we took out a motor, changed a motor on the water, the motor had to go on a barge. The barge had to be placed under the wing between the float and the hull so it didn't damage either, while the old motor was lowered to the barge and the other one was brought up.

Observant Gladys Ritchie watched the operations from her office.

They had wooden planks. Not being an engineer I'm not quite sure how they were attached to the flying-boat but I know the men had to work out there on these wooden planks which were attached to the wings of the aircraft, and do their work from there. At times it was very windy and these planks would be moving around and, of course, a lot of tools were lost overboard into the water as well. We didn't lose any workmen though.

Soon there was a slipway at Rose Bay for aircraft servicing. George Roberts worked on the first flying-boat to be beached.

The first one arrived almost to the top of the slipway when

Top: Gladys Ritchie remembered how men worked 'on wooden planks which were attached to the wings of the aircraft'. Bottom: Flying-boats at their moorings in Sydney's Rose Bay

the tractor towing it, with its rubber caterpillar equipment on it, started to slip in the water.

So the next thing the aircraft ran backwards down the slipway, taking the tractor with it into the harbour. I particularly recall this unflappable chief engineer, Arthur Baird, who stood on the top of the slipway and said, 'Well, we'll see something interesting shortly, won't we?' And there was the aircraft bolting backwards down into the harbour, the tractor going with it, and Arthur completely unflappable.

At the time the question was whether, once it was airborne, the service would meet passengers' heightened expectations. Captain Hugh 'Smokey' Birch recalls that in those early days it still took about ten days to get to London.

Later on it got down to about six. But every night they stopped somewhere and the passengers had a good night's sleep in a very nice hotel, such as Raffles or wherever. It was a very pleasant and very relaxing trip and there was certainly no such problem as jet-lag.

The flying-boat was built for water. Later, when pilots went across to India, people would say, 'How did you get across?' Well, we had to land in lakes and in other places we had to land in rivers. There was no prospect of landing on land.

The flights were leisurely. So much so that flight and cabin crew would devise interesting ways of passing the time with passengers. Keith Calwell enjoyed this aspect of flying.

We used to send down a position report every half-hour at least and say where we were and where we would be and what to look for on the way. There was a card to pass. Several cards were written out and passed around. The captain or the first officer would spend a lot of the time with the passengers in the cabin talking to them and making friends with them and showing them what was happening and that sort of thing. First officers used to like nothing better than to be left on their own and the captains used to keep an eye on where they were because they could look out the window and see just how far off course they were. Of course if you were over the ocean they couldn't see anything but they could check the drifts.

Cabin crew also came with the new planes. Journalist John

Gibbs worked as a cabin steward in those early days, a job which he regarded as a real privilege.

So few people flew and they were fascinating people really. They were often much more relaxed on board, naturally, because if they had any stress or any problems they were pleased to relax and have a few gin and tonics.

The cabin crew had to help in some of the procedures. You had to wind a swivel on each engine to stop it from overheating when you were coming down and I wound one of the swivels the wrong way on my first trip. I was petrified because the engine started to smoke! That was a man-made emergency I'd say.

Today's 747 or Airbus can surmount most weather conditions but the flying-boats of the late 1930s had a relatively low ceiling and, when they struck bad weather, limited speed to fly out of it. John Gibbs recalls that if they struck turbulence

they would drop and creak and lurch and groan, very cumbersome, very slow. I think their top speed was about 150 knots, very noisy. Some of the boys used to climb in the wing sometimes among the mail bags and try to have a sleep, if you can imagine trying to sleep with four engines thundering around you. That's how they were—very slow, very noisy, but very reliable.

Bad weather was just something you accepted. Pilot Keith Calwell and his colleagues had no choice but to fly through the weather.

The normally aspirated engines would operate to about 14 000 perhaps 16 000 feet. But if there was no oxygen carried for the passengers we weren't supposed to go above 10 000 or 12 000, and sometimes we used to sneak up to 14 000, and on one occasion I went up to 16 000 for about two and a half, three hours, in between Bangkok and Penang in order to make a stage without a stop, and regain half a day's flying time. I called up the stewards and said did they want to go round and see that no-one was turning green at the gills, that no-one was struggling for breath or anything, and they were to do regular patrols.

I did this while I was co-pilot, and the captain was downstairs talking to people and he eventually came back up quite happily, having had a good meal, and he was sitting in my seat because I was in his. He was quite happy for quite some time and then suddenly he looked at the altimeter and he said, 'My god! What's happening here?' And I said, 'Oh well, I came up'. And he said, 'But they'll all be going green,' and I said, 'Well they're not because the steward's watching them and we can just make Penang. If we do we won't have to stop at the other islands and we'll be able to make the flight.' So he just quietened down and we made the stop and saved half a day.

A four day trip to Singapore was regarded as really a very fast trip. Taking off from Rose Bay in Sydney the aircraft had several stops. Number one was Brisbane and then to Gladstone and Townsville, all in one day. Day two was spent flying over north east Queensland, across the Gulf of Carpentaria, brief stops at Karumba and Groote Eylandt and finally Darwin. It was two days before the flying-boats left Australian land and Australian waters.

For cabin steward Dennis Egan these flights were enlivened by glimpses of life below.

The skippers always made a point of hugging the coastline and the passengers could see kangaroos and buffaloes which was very entertaining. The only trouble was they rushed from one side to the other.

For a pilot like Lennie Grey who had flown QANTAS's DH 86, and many other aircraft in the 'Rag, sticks and wire' category, there was no going back.

Every time you step up from a small aircraft to a larger one, you do get quite a feeling of stability, particularly in turbulence when you're coming in to land. I think that applied very much because over the water you don't generally get a lot of turbulence so it really made it much more pleasant. Also from a crewing point of view, it was a terrific leap forward because instead of doing all those chores we had to do before, we now had still the two pilots, but we had a radio officer to do all the radio, and he also did the bow work of mooring, picking up the ropes and things.

It wasn't long before the pilots created a distinctive aura around their particular aircraft and the skills required to fly them. As Hugh 'Smokey' Birch admitted,

> we regarded ourselves, flying-boat pilots, as almost a race apart. In fact we were known by land plane pilots as the 'flying-boat union'. We established purposely a kind of mythology about the difficulty of flying a flying-boat which was not really true, but one has to remember that every landing in a flying-boat was different.

So was take-off, as Keith Calwell found.

> If you're flying a boat and you really want to fly it instead of just dragging it across the water from Kingdom Come, you have to manoeuvre quite a little bit. The way to do that, of course, is to pull the stick right back hard, then push it right forward hard and rock it, and at the same time, judiciously and with feeling, apply a little bit of rudder to kick it up a little bit. But you just had to catch that little bit of two inches, or one inch, that you were gaining until you gradually rocked it up on to the surface of the water, and then, once it was on the surface, it would take off on its own.

For skipper Phil Mathiesen, the pilot, apart from being an airman,

> must be a competent seaman. He is involved with all sorts of sea conditions and sea hazards of reefs, obstructions, other moorings, other craft. So to my mind the flying-boat pilot was a special breed. He had to enjoy working out in the weather. It was quite different from just operating on a land airfield.

And for Keith Calwell, a trip didn't end when the engines stopped.

> With land planes you land, run the thing to the ramp or wherever you're going to stop, switch off the engine, ask someone to put the control locks on, and hand over to the engineer. But with a flying-boat that's only when you start to work because first of all you've got to attend to the re-fuelling before you can go away.
>
> That's the duty of the first officer and you've got to check all the gear that goes off and see that it's right; see that the moorings are properly secured, take responsibility for them,

Coolangatta *taking off from Rose Bay.*

put the control locks in. In other words, you do everything that is necessary for the aeroplane to be safe over the night and, if necessary, you sleep on board.

On one occasion, during a bad storm, he flew his flying-boat on its mooring.

I had the four engines running at idle and the aircraft left the water several times for a few seconds and I just kept it there and I was ready to open up. If the mooring strut went I would have just opened up and flown away to some other area.

As cabin steward John Gibbs remembers, it was also vital to load the aircraft correctly.

You had a forward hold and a hold in the tail and we had mail bags and freight in the wings etc, and if you didn't load the plane properly from the datum point, which was the central point of gravity, then you could have a lop-sided plane, or a tail-heavy plane, and it was quite dangerous because you had to land on water, and sometimes if you're landing on water and you've got a swell, you can have a four foot wave coming in and hitting the aircraft and if you hit that hard, you really were in trouble. That was why those pilots had to be good in those days.

Pilots always say that, in flying, there are two critical periods, getting up and getting down. Landing at Rose Bay at night after a long flight was not always easy. Captain Birch did it many times after the war when Sunderland flying-boats still operated between Lord Howe Island and Sydney.

There were no navigational aids at Rose Bay itself except a light on top of the hangar there. You had to be fairly skilful in getting the aircraft on to a flare path which is a series of kerosene flares on the water.

But for Keith Calwell there were compensations, if you got in just before sunset.

I think the approaches to Rose Bay were much nicer than the getting away because you were a bit busy when you were getting out. Coming in to Rose Bay in the late evening and seeing the harbour and the sun glistening on the whole of the harbour and lighting all the buildings up was really a magnificent sight, and also at night (we did quite a few night flights) we used to fly round over the city and everybody enjoyed it.

One unpaid member of QANTAS ground staff was utterly reliable in predicting the estimated time of aircraft arrival. This was a cattle dog, owned by the watchman, Bob Kay. According to Gladys Ritchie, this ground observer could tell well before the arrival of the aircraft that it was due and would rush up and down the waterfront barking madly. Shortly afterwards the expected plane would arrive, wings glinting in the evening light.

Landings were slightly hazardous on Saturday afternoons when weekend harbourside sailors could crowd the waterways, often hoping to see aircraft arrive. Department of Civil Aviation launches patrolled the landing route to separate aircraft and spectators. They were also on hand to make sure there were no bits of wood or debris on top of the water that might hole one of the planes.

As Captain Birch remembers, flying-boat landings were very much part of the Sydney harbour attraction until well after World War Two.

It was quite a thing to do to come down to see us taking off from Rose Bay, although most of the flights took place at night. Most of the arrivals were in daylight so there was always a good crowd of onlookers to see us, and it was fairly spectacular to watch a flying-boat land.

But the war was to put a temporary, and in time a permanent, end to the flying-boat era. Hostilities closed the routes and the war itself produced aircraft that would make even this wonder of the skies obsolete. The Empire flying-boat scene, for all its glamour, lasted less than two years.

In September 1939 England declared war on Germany and while the service continued with some reduction between Australia and Britain, the capitulation of France and Italian entry into the war closed the Mediterranean for the British and the joint Imperial Airways–QANTAS partnership.

At the Australian end of the operation, QANTAS dispersed its flying-boat fleet, sending some to war operations and using quite different aircraft on the other side of the continent where they were to play an invaluable role in transporting high security information and passengers vital to the war effort. But the Sydney based flying-boat operation, brief though it was, had brought a new era into being, an era which Gladys Ritchie was delighted to have had in her life.

It was a wonderful experience for me and I guess I saw an entirely different side of life. If I had gone in to the city and worked in a commercial office right from the start, and stayed there, I would have had no conception about aircraft or the technical side of life.

Top: Securing the Catalina's mooring rope. Bottom: QANTAS Catalinas at their Swan River base in May 1944. Rigel Star is on the water and Vega Star is in the background.

Cats on the Swan

It was at a time when things were fairly grim here. We were worried about the Japs moving down, and those Catalinas started about that period. They were a very beautiful thing to see landing on the Swan.

Edna Daw

In the distance I could see these lovely silvery white things on the water, and I remember asking Dad what they were, and he said, 'Oh they're flying-boats,' and I looked at these things, and they looked so graceful sitting on the water and just bobbing at their moorings—to a kid almost like great big birds.

Margery Carr

My most vivid memory is living in Bicton near Canning Highway, and these things coming over the house and making such a terrible roar, and flying so very low that the tiles used to rattle on the house, and things in the house would rattle and bounce around a bit and this incredible vibration. It was daylight I think, most of the time, because I used to race outside and watch them come over. That's how I know that they were so low. But it was only in subsequent years of course, that I learnt that they were probably so heavily laden with fuel.

Kevin McQuoid

THE QANTAS PERTH–CEYLON
CATALINA SERVICE 1943–45

*E*dna Daw, then a young married woman, and teenagers Margery Carr and Kevin McQuoid were just three out of thousands who saw the Cats on the Swan in wartime Perth.

With many others of their generation they recalled unique glimpses of the Catalina flying-boats, aircraft which were to dominate the broad estuary of the Swan River for much of World War Two.

The story of the QANTAS Catalina operation in Western Australia is one of the most remarkable stories in civil aviation history. As the war intensified the future for QANTAS looked bleak. Thanks to Italian entry into the war the Mediterranean no longer offered safe passage to Europe. Worse was to come. The Japanese drive into South East Asia and the fall of Singapore marked the end of attempts to fly a scheduled service between England and Australia and Japanese military success blocked many other civilian routes out of the country. QANTAS was to lose three of its Empire flying-boats in action, two with crew and passengers over Java and Timor and a third in the Japanese raid on Broome in March 1942.

With connections broken south and east of Singapore the British and Australian governments discussed the idea of using the Indian Ocean as a secure air route linking Perth and Koggala, an RAF base just over 100 kilometres south of Colombo. It would mean non-stop flights of 3500 miles, an extraordinary task for those times. The aim was to transport both key war personnel and important coded information on microfilm. Some armed forces mail also travelled via Catalina.

QANTAS was given a civilian contract to fly the route in both directions and thus acted as a vital agent between two war zones and two leaders, Churchill via Mountbatten on the Indian sub-continent and General MacArthur in Australia.

Initially the airline had survived with charter work for the American forces but its flying-boat fleet in eastern Australia had been seconded to the RAAF for essential war work. So the prospect of a civil aviation Perth–Ceylon schedule was to prove not only of great help to the war effort but also ensured QANTAS's role in the post-war world. The aircraft they were to use, American-built Catalinas, were to be supplied to Britain by lend-lease from the United States and, at the Australian end, flown by QANTAS.

QANTAS founder Hudson Fysh put his personal commitment behind the project, travelling himself on one memorable thirty-two hour flight, despite considerable opposition from Civil Aviation authorities.

When the Catalina scheme was first suggested, Mr A Corbett, Director General of Civil Aviation, told QANTAS that he could not countenance such an exercise. It would be sending people out to be murdered. Fortunately for QANTAS and its passengers Corbett's worst fears were not confirmed but his apprehension was understandable.

One of the first to fly the new route was navigator Jim Cowan, recruited by QANTAS from the RAAF.

Anything QANTAS asked for they were to be given, and if QANTAS asked could they have a loan of Jim Cowan, they were able to have Jim, especially as I was quite keen to be lent. We were pleased to be doing a job of work that had to be done and was unusual. It wasn't sort of humdrum. Silly to say it wasn't humdrum, of course, after sitting for more than twenty-four hours in the air!

QANTAS flying-boat operations on the Swan River began in June 1943 and continued until 1945. The sudden development of an aircraft base on the Swan River provided an exciting possibility especially for young men. Cliff Brown, who later went on to become an international airline captain, joined QANTAS at fifteen when the airline advertised for a junior worker.

I applied for the job and I was pretty small and they needed someone small to get up to lubricate the floats, because we had to crawl through the wings to do it. I was just the lucky one who got selected.

When I joined QANTAS the Cats had actually been operating for around about six weeks. I joined them when all they had was a kiosk on the foreshore at Nedlands at the end of the pier.

QANTAS had initially delivered Catalinas to the RAAF across

the Pacific, so they had some prior knowledge of these flying-boats, but until they transferred their operations to Perth they had little knowledge of the Indian Ocean. Navigator Jim Cowan and his fellow fliers had to ensure that the aircraft would fly the longest sea-route in the world.

> They stripped everything out of those jolly Cats till they just had all they needed and then could only carry one thousand pounds, and the schedule was twenty-eight hours plus or minus four. I did the thirty-two hour crossing with them. Thirty-two hours non-stop in an aeroplane is quite a long while.

In time the operation became known as the 'secret order of the double sunrise'. The Cat flew all day and through the night and passengers saw the sun rise before they reached the end of the journey, whether it was at Koggala or the Swan River.

Alan Foster, who joined QANTAS in 1943 as a driver and later worked as a mechanic, saw many a Swan River take-off.

> Everybody stood and watched and as soon as you saw the flume on the top of the water fly up from underneath the wing, you were able to identify the fact that he had just gunned the motors, and you watched until you saw that last little speck of water at the rear end of the main hull part with the ship, and then you said, 'He's off!'

The great quality of the American Catalinas was their range. In addition, compared with many British aeroplanes of the time, they were state-of-the-art aircraft. John Gunn believes that in many ways they helped QANTAS adjust to the idea of buying American aircraft in the years immediately after the war.

> They were spotlessly clean, they had electric stoves in them for heating water, they had four bunks, they didn't leak oil.

QANTAS was taking no chances with its new aircraft and the extraordinary demands of the new route. Any aircraft bound for Colombo was test flown before every flight, a rule that applied equally to pilots and mechanics like Alan Foster.

> If you'd worked on an engine, you had to fly that aircraft with the captain of the next flight to make sure it was right. He said, if it's good enough for you it's good enough for him, sort of thing, and you might do anything up to two hours'

flight with them. But they were great. You'd stooge around over Rottnest anything from 6000 to 8000 feet, fly up the coast and do slow approaches and once your aircraft was air-borne and the captain said your part in the affair was finished, you could go and sit in what they called the 'blisters' which is two cupolas, one each side, in the rear of the aircraft, throw these open and the whole world would be open to you down below. You didn't wear a harness, you just made sure you didn't fall out. You didn't get dragged out because the aircraft didn't fly fast enough, but you were open to the wide, wide world.

Cliff Brown, then servicing the Catalina test flights, saw each take-off as a challenge.

The danger in flying-boat operations was having to abort half way through take-off because the machine would get into a porpoising situation and could actually go into the water nose first. It depended on the state of the water, of course. If the water was glassy cover it was very hard to get unstuck. We used a patrol boat to break the water surface up when it was very, very smooth day, and quite often they'd find that they couldn't really get unstuck even then.

It often seemed an incredible amount of time for an aero-plane to become airborne and they were very loath to leave the water. You must appreciate that with a very heavily loaded aeroplane, getting off the water is one thing and then getting it flying is another thing.

Ailsa Pearn-Rowe was a civilian who, quite by chance, under-took the very last Catalina flight in 1945. She has vivid memo-ries of trying to get off the Swan's smooth, early morning surface.

Our pilot had devised a way of bumping, jumping the plane to gain height. I don't know how it was done but it was very effective, I'm told.

For Alan Foster there was nothing like landing or touching down on water,

especially when it was smooth. This is why they used to always have an early morning take-off because the river was at its nicest behaviour, and other aircraft weren't being turned round. Mind you, we'd only have a maximum of three aircraft here at one time because there'd be one probably in flight or just put down; there'd be one at the other end which would be its next time around. So we had a pretty full period. You started at eight o'clock providing there were no departures on the day. If there were departures you might even have to start at six-thirty in the morning, because take-offs had to be done before the wind was too high. Summer time you'd get an early easterly which was ideal. If it was a screaming easterly then there were problems.

To commence your take-off, first of all your take-off path had to be patrolled by a small marine tender. He checked the area to make sure there was no floating debris. I've seen it coming up from Fremantle—stevedores might have thrown over lumps of timber when they'd been shoring up, and that would finish up on the flight path of your aircraft. There was not a great deal of private craft on the water in those days. The length of the take-offs was a long one because they were hauling full tanks to give them sufficient fuel for, say, a head wind flight to Ceylon.

This is where the experienced skipper, you might say, nursed his aircraft. If he had still air, in other words not much lift or assistance from the atmosphere, he'd have his engines flat out and, of course, the internal combustion engine could only maintain that rpm for about two to two and a half minutes.

Edna Daw always held her breath for the Cats!

Quite often when they were leaving from here they were very heavily laden with petrol in particular, or with fuel, because they had to fly non-stop, and it took a very long run at times, and there were times when you sort of had your heart in your mouth and you thought, oh they're not going to make it! But they did.

Alan Foster can still recall his anticipation. Would the aircraft he'd helped test get up and fly?

When you were sitting on the shore watching an aircraft take off, you'd think those wing tips were going to go for ever and ever higher into the heavens without lifting that aircraft. From the start of his run, he'd have his engines flat out. The skipper would be really battling, and he and the

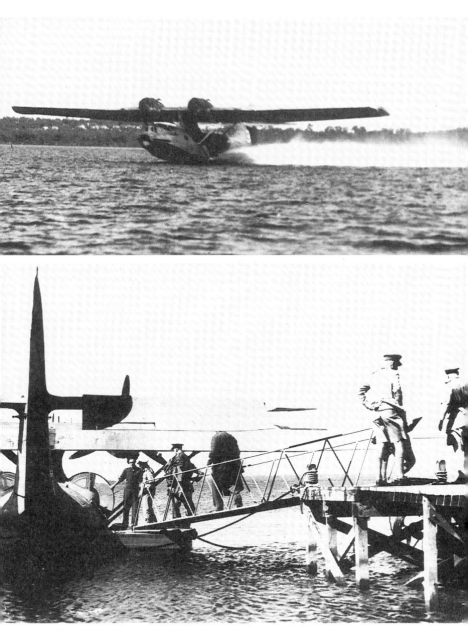

Top: A Catalina on the Swan River. Bottom: Passengers boarding a Catalina flying-boat for the non-stop trip from Perth to Ceylon.

first officer, once they could feel the aircraft starting to lift but not really taking to the air, he'd be hauling on his flaps with all his knowledge and keeping his fingers crossed at the same time, hoping that an engine didn't spit and he'd lose power and be back down in the water.

For Jim Cowan the biggest difference from a land-based aircraft was that

take-off runs were gi-normous. You virtually had to rock the thing back and forth to get it up on the step.

'Getting on the step' is an expression that has almost disappeared from modern aviation vocabulary, along with flying-boats themselves. Cliff Brown explained that getting on the step in a flying-boat

is a matter of getting on the forward portion of the hull so that there's less water resistance to the machine. Quite obviously, if you had all the hull in the water, you'd never lift off, so the idea was to push the stick forward, get it on to the step and leave it on the step until the aeroplane was ready to unstick.

Alan Foster remembers the aircraft as nearly always capacity loaded.

Sometimes they were really crammed. If there were passengers, VIPs to take, this made it more so because in flight they had to have some sustenance. It was all sandwich stuff and hot coffee which was pre-made, last to go on board.

The provisions were usually in large wicker hampers, according to Joy Mara, who worked in the QANTAS traffic office, but the people departing were more of a mystery.

They were bringing VIP personnel down and flying other people up who were quite important to our war effort. It was a very quiet, secret sort of thing and a lot of people had no idea that they were there for that particular reason. We never knew till we went in on a morning—sometimes it was six o'clock—to do the typing and things necessary who was there that we were meeting or seeing.

Jim Cowan recalls carrying generals, admirals, high-ranking officers or tech. sergeants who were the craftsmen.

But they had to sit in the seats at night whilst we got the bunks. Lie down rest was for the crew only, not for the passengers.

Aircrew consisted of two pilots, a radio officer, and an engineer. George Roberts reckoned the engineer fared worse than the two pilots.

He sat up in the pedestal between the hull and the wing where he could observe both motors from either side. His feet were in stirrups and he sat on this seat and how he maintained that position for hours I don't know. He would be in the noisiest part of the aircraft immediately underneath and just to the rear of the engine located in the main plane.

Alan Foster had, and still has, a high opinion of seaplane pilot skill.

They were so conditioned by their previous experience that they'd get out and say, 'Oh gee, it's bloody good to be back on the ground again'. The first step, of course, was not on the ground, it was on to the little marine tender that went alongside to take them off. But you didn't hear too many whinges from them. Sometimes they might have been in the air for anything up to twenty-seven hours.

Lennie Grey's previous flying-boat experience in peacetime Sydney was now paying off.

Fortunately, I think most of us who were initially involved in them had all taken on a little bit of seafaring acumen. And we did treat them as ships so we didn't find a great deal of difficulty in working watches on and watches off, and we carried sufficient crew to do this. We couldn't have done it otherwise.

They were very long flights, though, and in the main going from Perth to Ceylon was a warm flight, because we flew low there.

Those people with weather knowledge will understand that the trade winds are south east in that part of the Indian Ocean, right up to the inter-tropic front, and by staying down low we had the benefit of almost a direct tail wind. Coming back was an entirely different flight because we could not stay down in those strong south easters, and we had to go high.

The Catalinas were stripped of everything. They had no lining in them, no heating, no oxygen. Coming back we had to wear flying suits and we had to stay up high, maybe as much as 12 000 feet for maybe fifteen hours or more.

Jim Cowan's crew dressed in zipcoats and mittens and balaclavas.

We didn't have any heating capability because we were carrying so much in-board fuel that we didn't want to take any chances.

Pilot Lew Ambrose was highly conscious of the Indian Ocean as a war zone. He found the early flights rather frightening.

We knew that there was nothing we could do to save ourselves, especially in the first ten or eleven hours, but it's amazing how adjusted to this one becomes, and how in the end it was almost routine.

One problem for all aircrew was that no-one knew the exact range of Japanese aircraft operating to the north west of the Catalina route, flying out from bases on the conquered Indonesian archipelago, a fact that made Jim Cowan's task as navigator tricky at times.

We were on radio silence from two hours out of Perth to two hours into Ceylon. We used to get a mid-flight forecast because, bear in mind, if we were going to take twenty-four hours to fly it, by the time we got to the middle of that flight the forecast that we'd left on was probably twenty-four hours old. So they put out another forecast and they'd hammer it out three times without ever getting an answer from us. Our radio officer used to take down most of the message first time and fill in the gaps in the second two. But it was never acknowledged. Then he read the new forecast out to us and we'd re-flight-plan, and we never had a close shave, interestingly enough. We had one occasion, for instance, when a Cat by chance flew over Cocos in the middle of the night, because we used to pass Cocos at night, and we were supposed to avoid it because it was a prohibited area. However, every now and again you didn't know where you were so you flew right over it.

The Cocos Islands were right on the Catalina route to Ceylon and a vital part of the 'Double Sunrise' communications system. Lew Ambrose was well aware of the Japanese interest in the islands and how the Australians deceived them.

They used to come over and watch for us over Cocos but they weren't to know that there was an operating cable station still working in Cocos. They knew that there was a link in Java and they destroyed that but they weren't to know that there was also a link with Perth and, as a result of that, although we had to maintain radio silence all the time, the cable station was able to send messages to Perth which were then broadcast by various stations as though we were up in New Guinea somewhere.

Jim Cowan recalled one tense situation on the islands.

One of our aircraft force-landed there one night. They flew right over the Cocos first. But about an hour further on they ran into an engine problem. They weren't sure whether it was the instrument that was wrong or the engine that was wrong, but seeing as they knew where Cocos was, they went back, did a run across the lagoon, dropping flame floats, turned at the far end and landed along their own flame floats at night. Now, when they were on the water in Cocos, the Japs actually came over and got as big a shock as our fellows did, went away and came back at about three thousand feet. By this time our fellows had got off the Cat.

They dropped a stick of bombs across the Cat and didn't hit it. If they'd known they could have come down and just sunk it with gun fire because there was no protection. But they obviously got as big a fright as everybody else! So that was the only close shave I suppose that we had with the Japs.

Luckily, ninety-nine per cent of the two year operation was routine, routine that seemed natural to the men who performed it, but perhaps they made the Perth–Ceylon run look easy.

Jim Cowan still chuckles over the time others tried it.

The RAF in Ceylon where we landed, because we came in

and out so regularly, thought it must be a piece of cake. When they were called in on one occasion to help us out they short landed on the west side of Exmouth Gulf Peninsula, out in the open sea because they'd made such a balls of it. They reckoned that if QANTAS could do it, anybody could do it. In fact they found out they couldn't. They short landed on a flight that we went right through on the same day. When we landed in Perth everybody was very worried. 'What had happened to the RAF Cat?' And we said we didn't know, we hadn't even heard of them. We just kept on ploughing through; we went right to Perth whereas they didn't even get to Exmouth.

The QANTAS crews themselves navigated the Indian Ocean with a combination of celestial navigation by night and drift meters by day. Jim Cowan found there was little let-up.

The navigator had quite a lot of work and furthermore you got pretty good at it. You'd shoot a star through a hole in the cloud and work out afterwards what the star was because you couldn't afford to miss.

First officer EH Neal made one flight which brought home the importance of accurate navigation. It was the delivery flight of the fifth Catalina from Koggala. The Cat flew at 12 000 feet for over fifteen hours, took a bearing for Perth over Cape Inscription just south west of Carnarvon and started a descent through dense cloud to 2000 feet. At this level the cloud thinned but there was still no sign of land. Frequent attempts to establish radio contact proved fruitless.

A concerned Captain Ambrose then turned the Cat hard left, tracked directly east and after an hour sighted the west Australian coast some eighty kilometres north of the Swan River. Later investigation revealed that sextant sightings had been calculated using figures from the Air Almanac for the wrong date. The crew had in fact flown close to the Japanese-held Christmas Island on their way south.

•

The QANTAS Catalinas were not the only Cats on the Swan. Not far from their base was the 'Black Cat Squadron', US

Catalinas withdrawn from Manila after the fall of Corregidor. Their work, anti-submarine patrolling, was as secret in its nature as the Double Sunrise Operation. To casual observers there was little difference. But the Black Cat Squadron was well named. Their aircraft were, to quote one impression 'a dark bluish black'. QANTAS Catalinas were also camouflaged, painted in the tropical camouflage style of the RAAF.

Since both were using the same type of aircraft, contact was inevitable. Cliff Brown valued the co-operation between the two quite different units.

> We relied largely on a lot of help from the Americans who were just around the bay in Crawley from us. We had no tie-up with them as such, but it was just a bit of their good-natured generosity that we all got on fairly well together, and they gave to us quite freely.

The generosity came in handy from the outset. Jim Cowan remembers that QANTAS was very short of spares when the base was being set up.

> In fact, when Norm Roberts first went there, I think he used to say he went there with a box of tools and a spare set of spark plugs, that's all he had.

Norman Roberts was the engineer sent to look after the Catalinas in Perth. As he later told his brother, George.

> The only spares they started off with were fifty spark plugs that Norm purchased on their way over to Perth via Melbourne.
>
> When he got there he had no means of installing those spark plugs in the two engines of the Catalina. So, to obtain a spanner, a necessary spanner, he went through to the then US Navy Air Force base to borrow one, changed the spark plugs and dutifully returned the spanner. The crew then flew to Ceylon and back.
>
> When they returned Norm went to borrow the spanner again but there was a bit of reluctance. Norm had foreseen this possibility so while in Ceylon he'd bought some ebony elephants. On his next trip to the US Navy Air Force base stores, he had in his pockets a couple of ebony elephants, so when he asked to borrow the spanner again he produced

these two elephants and popped them on to the counter. The reply he got immediately was, 'We'll keep the elephants, you can keep the spanner'.

•

The Catalina flying-boats plied the Indian Ocean route for just two years, from 1943 to 1945. Like their civilian predecessors, the Empire flying-boats based in Sydney, their era was brief but significant. Just as the Short Empire flying-boats were displaced by war the Catalinas in turn were displaced by the technology which the war itself delivered. Strategic demands in Europe had produced the long-range bomber and it was these aircraft and their civilian descendants which would change the post-war aviation world. More immediately they would replace the Catalinas. For Cat pilot Lennie Grey that change was gradual.

> Before we left Perth we also had Liberators running in parallel with the Catalinas and, in fact, I got used to changing from two engines on the water in a Catalina to four engines on the land in a Liberator. The Liberator couldn't go direct. It had to go up to Learmonth. We had four Liberators and five Catalinas, and we would have eventually withdrawn the Catalinas but things changed. The end of the war came pretty quickly.

Jim Cowan recalled over 200 crossings in Catalinas.

> But it wasn't until the Liberators came in and gave us an eighteen-hour crossing that seemed like heaven, and then the Lancs came in and gave us a sixteen-hour crossing which seemed like even better heaven, that we realised that the old Cat had been a pretty hard slog.

The Cats nonetheless had done an excellent job. They had flown largely in radio silence across a dangerous ocean relying on astro-navigation for two years without a single mishap. Sadly, four of the five Catalinas were later scuttled off Rottnest Island as part of the US–Australia lend-lease agreement. The fifth was destroyed off Sydney heads.

Forty years later, in 1983, a granite plinth was placed on the edge of the Swan River near the University of Western Australia. It is the only visible reminder of the QANTAS base between 1943 and 1945.

QANTAS had achieved a good record in those years, one which would stand the national airline well in the post-war world. That longest hop in civil aviation history would set the pattern of renewal for the post-war years when QANTAS would again operate a service to Europe, with the Kangaroo route. They had earned their right to become Australia's overseas airline in the post-war world.

Ailsa Pearn-Rowe made the very last Catalina flight to Colombo in July 1945, a flight she has never forgotten.

> Perth had had rain for about three weeks and there was such a backlog of people waiting to fly abroad that when I was offered a seat in a plane I gladly took it. We had a cup of soup and a sandwich.
>
> There was no comfort whatsoever. We didn't know the comfort that people know today in planes nor did we have the food, either. We had three passengers, only three of us, in little dickey seats sitting side by side, and opposite us were sleeping bunks for the crew. There was a crew of six.

As the Catalina rose over the Swan River and headed north, she became aware of an overpowering smell of fuel.

> The two gentlemen passengers with me had nicotine on their fingers I noticed, and we weren't even asked to give up cigarettes. I think we were told not to smoke but during the evening I was a little concerned that they might take cigarettes from their pockets without thinking—but they didn't.
>
> The smell of gasoline was overpowering and we weren't allowed to stretch our legs or move in the plane at all. At one time I walked out into the bubble to have a last look at the west coast of Australia and I was quickly told to return to my seat as I was displacing the load.
>
> I know now that we were flying at about 1000 feet, but to me, a passenger, I could just look out and see the waves below and it didn't seem anything like that.

The journey was not totally without incident

> On this flight one of our two engines caught on fire. I was looking out of the window when I saw the flames even before one of the officers came to waken the captain. The other two passengers seemed to be asleep, but I noticed how efficient everybody seemed to be and how calmly they seemed

to take it. They switched the engine off, and we just carried on and limped in to Koggala Bay quite safely, but it was quite an experience.

I thought it was an adventure. There was a little reception committee awaiting us and a little ceremony as this was the last flight of the Catalinas westward, and we were presented with certificates of the 'secret order of the double sunrise'.

For those who flew the Cats, like Jim Cowan, 'It was a good service and it was rather good in retrospect to have belonged to it'.

And, with the last word, Cliff Brown.

It was just a part of history and you could see it developing in front of your eyes. It was quite amazing really to have an appreciation for that sort of thing as it's happening, and knowing that it's going to be part of history and everyone knew it was.

Scraping Gravy off your Armpits

Flying over Shark Bay in a DC 3 at 6000 feet was the most fascinating thing that I've seen in the world. Then, during the wet season, you'd be flying through the Kimberley and when the rain was very, very heavy, I've seen the edge of the whole range there, pouring in like one waterfall and I don't think I'll ever see Niagara but I have seen a Niagara in the Kimberley.

Air hostess Margaret McDonald flying in the late 1940s.

POST-WAR FLYING IN
WESTERN AUSTRALIA

MMA hostess Margaret McDonald

M argaret McDonald joined MMA, MacRobertson Miller
Airlines, in April 1947. She was one of the earlier
post-war recruits for the Western Australian airline after World
War Two. The war had severely disrupted civil aviation but
there had been one bonus: after the war there was a supply of
both pilots and planes, a renewed demand for air services, and
a demand by adventurous women who'd been 'manpowered'
during the war and now, like their male counterparts, were
looking for more challenge in peacetime Australia.

This is the story of how they and their passengers fared in
those early years after the war in the huge and isolated western
third of Australia, an area with sparse and scattered population,
especially in the north west.

The Kimberley and the Pilbara still rely on scheduled
aircraft services, much as they have since the 1920s and 1930s
when Norman Brearley's Western Australian Airways first took
air mail into the remote north west.

In the immediate post-war period, however, the north-west
air schedules were run by MacRobertson Miller Airlines, who
had won the air mail tender from Brearley in 1934.

MMA had attempted to provide a regular service to the
north-west for as long as possible during the war but Japanese
bombing of Broome and Port Hedland restricted, although did
not totally suspend, their normal operations. For much of
the rest of the war MMA performed a valuable role in supply
and transport work.

When hostilities ended in 1945 it was time to renew normal
operations. Margaret McDonald recalls the atmosphere of
those early post-war times.

> You had a sense of the war still being very close to you. Pilots
> had just come back from the war and it seemed to be part of
> the adventure I'd missed by being a timid little bank clerk.
> I felt I'd really missed something in missing the war.

That might have been the fate of one pilot, Bill Anderson.

> When I came back from the war I rejoined a bank which
> I worked for and whilst I had good career prospects, I felt
> that aviation was really the thing for me. It gave me more

opportunity, I think, to express myself. There were a lot of advantages over the horizon at that time. It was a job which didn't pay particularly well, simply because there were so many pilots around and the supply and demand more or less dictated the salary. I had to actually drop salary to leave the bank to go flying, but I thought it was well worthwhile.

Margaret McDonald, whose career in flying was quite short—she only flew for three enjoyable years—nearly didn't make it into the air.

I suppose at the time I was being the first 'hippie' because as soon as I was un-manpowered from the bank I left with no idea of what I was about to do, and my mother was always cutting little bits out of *The West Australian* and telling me to apply for jobs. One of them was a ground hostess with MMA and she said to me, 'You know, you might even get a ride in a plane'. But they said no, they'd filled the position. Then a few months later they wrote to me and said, if I was interested, to come in and be interviewed. What had happened was that the hostess they had appointed spent the night of every day that she flew in a hospital somewhere on the coast; she really suffered terribly from air-sickness. I often say that they probably chose me because the uniform that she had fitted me. But nevertheless I went off there and I was thrilled to bits.

Hilary Fox had done a man's job during the war, also working in a bank. With the return of servicemen to their old jobs she was unwilling to take a less interesting position.

I had my heart set on something quite different, and something quite different was being an hostess. So I went down to see Cyril Gare who was the general manager at MMA at that time, and he said, 'Well yes, Hilary, I need a private secretary. You can work in my office until such time as an opening comes up in the airways.' It was so hard to get out of that private secretary's job into the flying.

Every time an opportunity could have arisen there was a good reason why I should stay in the office. But I stood out for it because I needed to do this something different in my life, and that's how I finally got to be flying.

Training, which today is a key part of airline operations, was then very much a matter of individual initiative. Bill Anderson spent four weeks of his own time studying engineering and flight training on the DC 3. Recently recruited Hilary Fox had no training whatsoever.

They just said, 'Well, righto, you're now the new hostess. There's your flight. You'll be picked up at four-thirty'. We used to fly the three days straight flying: Perth to Broome; Broom–Darwin–Broome; and then Broome to Perth. I often used to think as I was walking down the street, well I'd be the only one apart from the pilots who'd been all the way to Darwin and back in the last three days.

Young women who became air hostesses were often seen as glamorous. The uniforms certainly projected a smart image, with their wartime wings look, gold buttons and sharp lines. Pat Jordan whose family lived in Broome, joined MMA from the north west itself. She wasn't too sure about the uniforms.

Looking back on it, compared with what flight attendants wear now, with designer label outfits, it was probably rather terrible. It was a khaki drab-coloured skirt and shirt and a maroon tie, and over that you had a coat with about four pockets and brass buttons and epaulettes and lots of brass to polish, including these giant double wings which in flight, when you bent over to do something for a passenger, used to stab you in the boobs.

When Margaret McDonald began work there was one air hostess and twenty-one passengers.

It was possible to serve breakfast between Perth and Geraldton—one hour and thirty-five minutes—with twenty-one passengers, but you had to get a tray out for each person and put everything on it: a bowl of Weeties and a tumblerful of tomato juice and coffee or tea and bread and butter and marmalade. I think I had a lifelong hatred of marmalade because you used to have to put it in little plastic dishes, and cleaning up afterwards and getting the marmalade off the dishes, it would get on me and my elbows …

Pat Jordan served salads with cold meat.

For that time I think they were very well-presented, but the regular travellers, like commercial travellers and the flight crew, thought they were rather like rabbit food. They did try to introduce hot meals at one stage between Derby and Wyndham, which was highly unsuccessful. It was a very rough sector going over the ranges, in the wet season particularly, and no-one wanted a hot meal. A lot of them were doing other things. They served them in tall urns and by the time you got down halfway you were scraping gravy off your armpits. So that didn't last long—back to salads and cold meat.

Margaret McDonald felt that food, much as it does now, filled in time as well as stomachs and sometimes eased the queasy feeling of air-sickness. But there was one flight where the food didn't go down well at all.

A friend, Joyce Vale, had a missionary and his wife on board for one spectacular landing on this particular stretch between Carnarvon and Port Hedland and over the Hamersleys. She went around and she picked up twenty-one bags full of, well, the lunch, and she had said to the cleaner, 'I think you'd better take the airsick container and empty it behind the hangar, or whatever you do with it'. Then the wife of the missionary came up to her and said, 'What have you done with my husband's teeth?' She's a very dry sort of person and she said, 'They're behind the hangar over there and you can go and sort them out yourself!' She said it was a very, very sad sight, the two missionaries over behind the hangar.

Check Captain Syd Goddard recalls airsickness as an every day, almost every flight, occurrence. Flights were generally bumpy because of the inability of the DC 3 to fly above turbulence levels—normally below 9000 feet and frequently quite low if headwinds were strong. So it wasn't at all unusual for most of the passengers to be airsick,

and in the confined space of those small aeroplanes, a few people being ill virtually set the others off and made it very unpleasant for them.

Terrain and climate didn't help. Bill Anderson reckons the area round the Hamersley Ranges to be some of the roughest flying conditions in the whole of Australia.

In that area you were subject to a lot of cloud flying in the wet season, particularly in the north-west. You had no radio on the aeroplane so, once you got into cloud, if there were thunderstorms about, the prospect of penetrating a storm was fairly real. Also in the low altitudes, of course, you get a lot of turbulence, even clear air turbulence.

Hostess Margaret McDonald suffered badly for the first three months.

> I would cope with everything during the day. It took three days to go to Darwin and back and when I relaxed—oh no, not again—I would retire and politely fill a little bag and dispose of it and come and sit down again smiling, as we put our wheels down at Broome (we overnighted in Broome at that stage). That lasted three months and after that I don't think anything ever happened.

Washing up for twenty-one passengers in turbulent conditions while descending towards a north west airport and seeing the washing-up dish leaping around in the buffet could be, to quote Margaret McDonald, quite exciting. Pat Jordan had similar problems.

> Sometimes the buffet was near the toilet at the back, and you had this wooden cupboard containing all the plates and so on which you put away, and on a rather steep descent sometimes the cupboard would fly open and all the plates would shoot out down the aisle.

Despite the rough ride, cabin and flight crew alike had great confidence in their aircraft. Apart from the Lockheed 10, MMA flew war surplus Ansons and DC 3s, an aircraft which Captain Harold Rowell felt was part of the family.

> The DC 3 I always looked on as a kindly old grandmother. She was a very docile aircraft and did a wonderful job. It has been said that it was so good that it retarded the advancement in aviation. They were plentiful and cheap after the war and we were able to give a much better service as a result of them being so cheap and amiable to landing on gravel strips and elsewhere.

Rowell's family feeling was shared by Pat Jordan.

> Even if I see it in an old film now there's always nostalgia … you always felt as though it was a very reliable friendly aircraft. If you'd weathered a few tropical storms in it, and

Margaret McDonald carrying boiling water on board for the tea and coffee and, if the water held out, for washing up.

it had been bogged and you'd sometimes flown in it with one motor feathered, you felt as though it would survive anything.

Margaret McDonald thought of DC 3s as not only tough but lovely.

I think anybody who's ever flown in a DC 3 will love it until the day they die, because it is the most beautiful plane. It's a classic shape. It's a reliable aircraft. It was extremely comfortable to fly in. The DC 3 was the nearest thing to a bird.

Proof of the DC 3's reliability was not something pilots chose to test unless they had to. Inadvertently Margaret McDonald was to provoke just such a test.

The first time I was up in the cockpit I said to a pilot, Reg Bagwell, 'What does it mean to "feather" a motor?' He said, 'Oh haven't you seen a motor feathered?' And I said, 'No, what does it mean?' And he said, 'Well, keep watching that starboard motor and I'll feather it'. So I looked out there and to my horror I saw the engine stop, and I said, 'Unfeather it! Unfeather it! Get it going again!' And he tried to but it wouldn't unfeather. That was my first one-engine landing, so I knew how very, very good the DC 3 was at one-engine landings right from about the second week flying.

Bill Anderson took off in a DC 3 from a station near Wittenoom Gorge on a very hot February day.

The aeroplane was loaded right to the plimsolls. The temperatures on the engines were very high and as we took off we got ourselves airborne and everyone breathed a sigh of relief. We were watching in particular low oil pressure situation with a high oil temperature, which is a bad situation. The hostess arrived in the cockpit and she said, 'You have an engine on fire'. Now in those days there was no fire detection equipment and there was only one fire bottle carried. So the engine which we were concerned about was observed visually and the answer was, 'No, the engine's not on fire'. I happened to look out the left-hand window and the flames from the engine were now back past the main cabin door.

Anyway, to cut a long story short, we did all the correcting drills and shut the engine down and got ourselves back on to the ground, and the station owner came up to meet us and he said, 'Oh I thought you'd come back,' he said, 'because just after you left I could see you'd caught fire and I expected you to come back.' I always remember one chap who said, 'Well I thought that happened every time.' If you can imagine a wall of flames about four feet wide and about twenty or thirty feet long going down the side of an aeroplane, it would be fairly exciting on any aeroplane that took off, to see that happen.

Margaret McDonald came into Port Hedland one afternoon at the end of a rough trip.

It was latish in the afternoon, and I'd asked one of the DCA men to keep an eye on the passengers. 'I'm going to have three minutes' sleep and I'll be right,' which I could always do. As I was having that sleep I heard a most terrible noise and I thought, good heavens, all the cans of fruit have fallen down in the buffet, but then I had this frightened man looking in my face and saying, 'The engine cowling just went past the window!' I said, 'Don't be ridiculous!' Then I looked out and there was a large smear of oil going over the wing. Actually it had stopped pouring out but because of the wind passing over it looked as if gallons were flying over it.

So anyway I got myself together and cheerfully smiled at everybody and went up front and said to the pilot, 'Did you know you've lost an engine cowling?' They said, 'Yes, do you know we've lost a pot (which is a cylinder)'. I said, 'Oh no, I don't believe it,' and I leant over the first officer and had a look out. It's the most horrifying thing to see a cylinder missing there. It's like when you've had a tooth out and you wake up the next morning and your tongue runs over and you say, oh no, what's happened? So I swooned away and then they said, 'Go back and tell the passengers we'll probably have to stay in Port Hedland', So I explained to one lady there, who was Matron Howell from Derby, that we'd probably have to stay in Port Hedland overnight, and that we'd lost the engine cowling, and she said, 'Oh, I have to be back in Derby tonight'. She said, 'Don't they carry a spare?' Everybody took it very calmly, especially the pilots because the DC 3 was absolutely wonderful for one-engined landings. You feather a motor there and the pilot would say, if they had to, 'You are now going to make the best landing in the whole of your life', and indeed they did, but you could never taxi with just the one engine.

The sequel to that story was that MMA Managing Director Horrie Miller flew into Carnarvon from Broome and quizzed Margaret about the likely location of the missing cowling. Could she remember where they were when it fell off? Later it was spotted from the air and retrieved.

Fortunately, incidents of this kind were rare and in those days captains like Syd Goddard had the time to hearten anxious fliers.

A lot of passengers would be on their first flight, some people feeling very nervous about it: the noise, lack of air-conditioning. Sometimes it was hot, and to be able to walk down and to reassure people and just have a friendly word here and there seemed to be a good idea.

However, Margaret McDonald was unable to persuade one passenger to stay aboard.

He came out to get on the aircraft three times. I shut the door and then he rose from his seat and raced back again and I opened the door and I let him out, and on three occasions he attempted to fly. Then the last I heard of him he got in his car and he was driving to Roebourne, and that wasn't easy in those days because the roads were very, very bad.

If flying was the way to go, the passengers needed every encouragement. Margaret McDonald's work began, while the aircraft was still on the tarmac, walking up the steeply sloping DC 3 floor, handing out newspapers.

In those days we handed out barley sugar, too, for the ears. And I did find one passenger at one stage who, when I said, 'Well, here's the barley sugar,' said, 'Oh thank you very much. What is it for?' And I said, 'Oh it's for your ears'. I came back to find that he had in fact put them in his ears.

There was plenty of time to get to know the passengers, whether they were novices or old hands. On the more remote north-west stops, such as Port Hedland, which today boasts a smart new terminal, passengers waited in a bough shed or, if such luxury didn't exist, under the wing of the aircraft which provided shade from the sun's heat or the 'wet's' downpour. It was certainly a way for crew and passengers to get to know each other well. For Hilary Fox that was one of the rewards of the job.

We would get on the planes in the morning. We would know our crew, we would know our captain and co-pilot, not only because we had flown with them before, but we had seen them in the office, we'd had cups of morning tea with them after our return flights, we had slept in the same hotels on the previous trips. We knew them, and we knew the passengers. They were not tourists, they were people who lived in the Kimberley, down on holidays, down on business, coming to work up here: the Shell man or the BP man. It was the same man every year. We'd say, 'Oh, so-and-so is on the plane again'. School holiday time we'd know all the children. It was a whole world apart from what it is today.

And picking up the passengers for the next leg of the flight was easy.

Station people usually stayed at the hotel. Therefore they were probably at the same hotel where we were, and we'd just say we were picking up five, and we'd make sure that we found the five in the hotel. But if a person was coming from Broome or from Derby, well we would quite often go round to their home because the towns were not that big that we couldn't find their houses, and we knew them anyway.

Knowing your passengers so well often meant extra but ungrudged labour for Pat Jordan.

Sometimes the short time you did have off in Perth—because we flew long hours, you know—was spent shopping for people up there: buying race trophies, and gifts for blokes for their girlfriends. It was all a happy arrangement; no-one was complaining.

And Hilary Fox could never understand how

women expected young hosties, twenty-one years old, to choose for a woman of, say, forty, a hat or clothing. 'Buy me a pair of shoes. I take a size eight.' But they did, they trusted us and whatever we bought was what they wanted.

Margaret McDonald felt that, while she was an occasional

Top: Pilot Sturdee Jordan and hostess Pat Jordan. Bottom: A DC 3 with aircrew and passengers in the north-west. Pat Jordan is third from right.

'drop-in out of the sky', it was her passengers who faced hardship and isolation.

> You became great friends with everybody in the north west because they were isolated. You saw the people who were working on stations and station owners' wives and governesses and children. You saw them as having the hardships. We had contact with Perth and quite often we would bring green vegetables up to them, personal little parcels, and shoes; and a friend of mine tells everybody that I saved her from insanity in Derby because I used to bring her books up there. I think the cockroaches must have eaten all the books in Derby at that stage.

Hilary Fox was to appreciate the benefits of aviation, literally from below. When she left MMA she married a station-owner and lived the nor'west life herself.

> Every time the station plane flew over our little station on its way from Derby through Noonkanbah where it had landed, to its next landing ground at Fitzroy, they would throw the daily paper out to us over our windmill. We would rush to the mill to get it and they would pride themselves on their accuracy of bombing! Inside the paper was usually a little note or a letter or something, some item of news from MMA that they knew we'd be interested in. So it was a grand day when the station plane flew over.

In Margaret McDonald's three years with MMA she became aware of the range of developing and changing industries in the north-west.

> One of them was the whaling industry. I had a plane load of Norwegians. They were quite incredible, extraordinarily large men who spoke no English and ate everything on the aircraft. I thought they'd finish up eating me.

The north-west was and still is notoriously thirsty, a factor Pat Jordan took in her stride.

> We used to take loads of shearers up and Cockatoo Island workers. Very often if they were coming south they would have had a bit too much to drink, and they also had hip flasks, and we had to frisk them, you know, rather nicely (they were rather good-natured about it) because you couldn't

have them drinking on the aircraft when they were in that state to begin with.

Margaret McDonald knew the sobriety standard for intending passengers. If they could walk up the aircraft steps they were sober enough to be flying with the airline,

> and of course the marvellous thing about it was that we didn't serve alcohol on board the aircraft, although sometimes passengers would sneak it on; and to my surprise one day I saw two pilots horror-stricken. They'd taken a bottle off a man and it was Shell lighter and cleaning fluid and he was having a bit of a gargle with that. That was about the worst that ever happened.

The north-west was also a place where you went to start afresh, evade a difficult commitment or keep out of trouble down south. Syd Goddard flew aeroplanes full of workers for Cockatoo Island and a lot of other places.

> In some of the cases we had more Smiths and Browns and Jones on board because a lot of those fellows were leaving home for various reasons, and in a lot of cases you'd land at Derby or Wyndham and the policeman would be there when you got out. Now he wasn't there to arrest anybody, but he just wanted to know who'd come in to his town. He'd have a look at the passenger list but it didn't tell him very much. That was the type of passenger we were taking.

MMA not only flew the north-west routes but across to the Northern Territory. In the 1940s the lengthy Darwin–Perth connection was the only way returning sandgropers and European migrants could reach Western Australia. Pat Jordan had some sympathy for the newcomers.

> There were loads of Greek migrants. None of them spoke English and they were all very anxious obviously to get to Perth after their experiences in Darwin and, with our kangaroo hop all the way to Perth, at Wyndham they'd all get up and gather their belongings and yell out, 'Perth!'. We'd have to physically manhandle them back on to the aircraft and so it would go on, you know, Derby, Broome, Port Hedland. They felt it was a myth—Perth was a myth.

Today you can fly from Perth to Broome by jet in three hours and reach Darwin in three and a half. In the DC 3 days the flights would stop everywhere and flew low and slow. There were over 130 airports throughout Western Australia and the Northern Territory, some of them remote station strips, others, at coastal towns, more sophisticated but still sometimes with bough sheds for terminals.

The complexity of those routes, compared to today's scheduled jet service to ten airports in the north-west, provided one great social and physical benefit, not to say peace of mind, for the scattered population of the vast nor'west.

Captain Bill Rowell, who joined MMA in 1945, working first as a traffic officer in Derby, pointed out that MMA had the contract to operate the Flying Doctor aircraft out of Wyndham, Derby and Hedland.

> Initially I would say that the Flying Doctor Service got precedence. Early on the MMA service was a very local one, a very domestic one, and I would think the potential customers for the Royal Flying Doctor Service were also customers of MMA, and anyone who was delayed for a Flying Doctor emergency flight would never complain. They would understand the situation and would support us in giving an emergency medical case priority.

The flights also involved long stop-overs. Margaret McDonald would sometimes spend two straight days in Derby during the 'wet'.

> And during the 'wet' the same things happened the whole time. The young stockman would come in. He'd buy his ticket to Perth, he'd go to the bar and wouldn't leave it until his cheque ran out. He would never get to Perth. Then he would go into the horrors. The Port Hotel, Derby, during the 'wet' was not the most pleasant place.

Today, for an aircrew, a round trip to Derby or Broome, including flight planning and post-flight sign-off, can take less than eight hours. In the 1940s a trip to the north-west meant days away from home. Syd Goddard and his contemporaries in the 1940s and 1950s not only flew the aircraft and did all the paperwork,

Top: *Refuelling in Western Australia's north-west in 1963.*
Bottom: *Flying Doctor JJ Elphinstone seeing patients at Forrest River Mission in 1948. His Lockheed Electra is in the background.*

but often did their own re-fuelling on the ground and loaded and unloaded luggage. They also spent a lot of time away from home.

Many of us who had families were accused of not being around to bring the children up. But the modern pilot joining the airlines today just couldn't believe the conditions. In places like Port Hedland and Derby, most of the time we'd be sleeping on one of those army-type camp stretcher beds on the verandah, and in the morning there'd be a salt water shower. They supplied that—I think it's 'Seagull'—soap to have a shower, and that meant tipping the bucket of this brackish water over you for your shower. Even when we used to fly through to Darwin, and stay at the Darwin Hotel it was thought to be quite a luxury, but even there there were four to a room without any space even to hang your clothes.

The discomfort was shared by the cabin crew. In remote northwest pubs Pat Jordan usually shared a room with someone else.

And sometimes if there wasn't room you'd have a bed on the verandah and in the early hours of the morning when you got up you'd have to trip over stockmen's boots and piles of trousers on your way to the bathroom.

Margaret McDonald knew she and others were roughing it.

You worked hard. I used to get up at quarter to three in Perth and rarely got to bed before eleven or twelve and then you got up the next morning at about four o'clock and started flying at five. I used to get very confused when we stayed overnight in Darwin, because there was the one and a half hours' difference. I never knew whether I had an extra hour and a half or I didn't, and I'd find myself coming home from a party to meet the rest of the crew coming the other way, and I'd made a mistake once again—that I didn't have that hour and a half.

Hard work! Well paid? Pat Jordan earned six pounds a week which she considered a good wage for those days.

We flew a lot of hours because there was no union restricting you so they could just fly you as much as they liked. There were only six of us when I was flying and in three and a half

Top: The old MMA hostel at Derby in 1950. Bottom: The Port Hotel Derby where many aircrew stayed.

years I logged three and a half thousand hours which I think by comparison these days is rather a lot.

Two final reflections on the DC 3 days of the 1940s and 1950s from former pilots. First Captain Harold Rowell who flew with MMA between 1945 and 1981.

When I joined MMA I was number fifteen in the pilots' strength. We did get as high as 135, so the progress in aviation is considerable in that fifteen of us used to run around with ten passengers at 140 miles an hour; subsequently still only two pilots running round with sixty-odd passengers at 400 or 500 miles an hour.

Former chief pilot Alec Whitham now lives in Shark Bay in the north west. He has seen enormous changes in aviation. In the late 1930s he flew the seven-day round trip Perth–Daly Waters link between QANTAS and MMA passengers. After the war he flew thousands of air miles in the north-west and saw the way the aeroplane changed this remote part of Australia.

Well, in the actual north-west itself, especially in the wet season, we were the only form of transport. There was no other transport except by packhorse and you can imagine how long it would take a packhorse to go from Wyndham to Halls Creek. I really think without aviation the north-west would never have opened up as fast as it did.

Struggling into a Fresh Clean Uniform

I returned to a rapidly expanding airline with a great spirit, that was filled with a lot of new people coming in who'd come from the Air Force, or from the services, or even elsewhere. New aircraft were coming in.

John Fysh after his wartime military service.

We always had to change our uniforms and I remember they were white sharkskin. Struggling into a fresh clean uniform just before we arrived, in those tiny little loos on the DC 4s—honestly, there were times I just passed out and then came to and struggled on. So there were very many quite tough aspects to flying.

QANTAS air hostess Margaret Carter.

NEW FACES NEW PLANES
QANTAS 1945–1980

W*hen World War Two* came to an end in 1945 a new era opened for aviation around the world. There were new planes, born of the war technology, new and longer routes because of improvements in aircraft design and performance. There would be new faces in the cockpit, and in the passenger cabin. Before long there would be a new kind of passenger.

These changes affected Australia in much the same way as they did Europe and America, but with one important difference. We were a buyer, not a maker, of aircraft. Traditionally Australia had purchased planes from British companies like de Havilland and Short Brothers, and the pre-war partnership with Imperial Airways had helped ensure a uniformity of aircraft between Britain and Australia. The QANTAS wartime use of American Catalinas had been an exception that became a precedent for change. Australia needed aircraft quickly if civil aviation was to revive. But, as aviation historian John Gunn points out, this wasn't easy.

> The aeroplanes that were possible, in the immediate post-war world for Australia, were extremely limited. We didn't have access to American transport aeroplanes. In fact, if I remember rightly, in the middle of the war they did a count. Australia had about twelve obsolete transport aeroplanes; America, in our region, had a total of over 450 relatively modern transport aeroplanes. That was the kind of competition, but we couldn't get at those aeroplanes because of 'lend-lease'. But at the end of the war, America was the only nation in the west who had these long-range transport aircraft.
>
> What you were going to see after the war was an abandonment of the old DH 86 type plywood aeroplane, and the flying-boat. You were going into a completely new era that was going to be dominated by America.

Coupled with an aircraft shortage was the question of QANTAS's future. In Britain a post-war Labour government had nationalised the British Overseas Aircraft Corporation; KLM in Holland was government owned; and under the Chifley Labor government Australia would follow suit, buying out the British half of QANTAS and the private Australian stock remainder. It was an amicable arrangement for both parties. But Hudson Fysh recalls, there was one divisive issue.

> Well, did we have a fight to have the name QANTAS retained! Were there any plans to have the airline called 'Air Australia' or anything like that? There was a suggestion. I remember I fought it pretty hard and my board were pretty keen on keeping the name on. But it is unthinkable that the name QANTAS should be changed—quite unthinkable. I remember we got past it by putting in very big letters on our planes, 'QANTAS—Australia's Overseas Airline'.

A battle that had to be fought and won again recently. But a key question remained. What aircraft were available for QANTAS to fly? And were they to be British or American? QANTAS chief navigator Jim Cowan knew Hudson Fysh, company founder and now chairman of the new nationalised QANTAS, to be a very strong loyalist, or royalist.

> On the other hand he was a very pragmatic man, and he had always recognised the value of the American aeroplane. In fact, he went across to pick up the DC 3 initially. Here were metal aeroplanes when Britain was still building string and paper bag. They were always a little bit behind. The problem with Britain, looking back now, was they they knew they were good and thought they were the best.

But within Australia, despite the massive shift in world power, brought about by the war, pro-British sentiment was still strong. There was always the feeling that when Australia didn't buy British aeroplanes it was somehow ratting on the Empire. The newspapers still advocated buying British but Britain wasn't building suitable aircraft for Australia's vast international routes.

QANTAS engineer, Ernest Aldis, was in no doubt as to which way to go.

> After doing our sums we realised that we just had to buy American aircraft. Under great pressure, the Civil Aviation Department finally agreed that we could import American aircraft and put them on the Australian register.
>
> Strangely enough, every American aircraft QANTAS bought, BOAC—in spite of having to take British-built air-craft to keep the British aircraft factories going—in every

case they bought similar to the American aircraft that we did.

John Fysh was in London when that decision was made.

> It was understood why we'd done it, but I think there was a resentment. The resentment was there probably at the higher levels and certainly in pubs and bars when I went to England. If you were an Australian you would be accosted for not buying British. But it was an emotional thing. It wasn't a rational thing, so one just had to pass it off. Fortunately, the Prime Minister then, Ben Chifley, had the good sense to withstand that pressure from England, and to approve QANTAS getting from the Empire dollar pool the dollars necessary to buy it.

In pilot Lennie Grey's view that decision ensured the survival of QANTAS.

> If QANTAS had had to buy I think it was the Avro Tudor, or whatever other British aircraft was available at the time, all of which tended to fail, I don't think QANTAS would have survived.

Australia was very aware that America had shown the lead with its DC 3s and DC 4s. Both these aircraft were to be used extensively by domestic and international services in Australia. But the aircraft QANTAS had its eye on from the outset was the Lockheed Constellation, a pressurised four-engined airliner developed during the war with a view to peacetime expansion of passenger traffic. It was an aeroplane which would put QANTAS on its feet as an international airline. Chief navigator Jim Cowan went to the United States to help check out the new plane.

> I went on all the first flights. It's interesting to look at an aeroplane that's rolled off a production line, on the end of the runway, and wonder whether it's going to fly. The thing has never been in the air. You know when they push a boat in the water, at least the damn thing floats or sinks. But an aeroplane, you don't know whether it's going to fly until it's at least half way along the runway.

Cabin crew reaction to the new slim tri-tailed aircraft was positive. Dennis Liston sums up the change.

I guess the simplest way to illustrate it is to say that flying to New Guinea from Australia in the early 1950s was on a DC 3, and flying to London was on a Constellation. Everyone knows the DC 3—un-pressurised, cold food, one flight attendant, one toilet, pretty bumpy ride, twin-engined aeroplane with the obvious things. But the Constellation ... pressurised aircraft, a lot higher altitude, toilets for men and women in those days, separate toilets, a purpose-built galley designed and built for the aircraft, a good hot water system, a good galley system really with hot and cold running water, and ovens—electric ovens. We had ovens on the DC 4s and the Constellations. We didn't on the DC 3s and the flying-boats—that was flask food. It came on in flasks and was always cold by the time you ate it. On the flying-boats and the 4s and the Constellation, we were literally opening cans to get the meals ready. Always beetroot of course. When you made a salad you had to have beetroot being an Australian airline. It was compulsory.

With or without beetroot, air hostess Margaret Carter was an instant fan of the new plane.

Well the Connies were, I think, the most beautiful looking aircraft that have ever flown the heavens. Just the most beautiful aircraft, and so it was a great step up. It was very glamorous and wonderful.

The hostesses themselves were part of the glamour of post-war flying. Before World War Two, female employment in aviation was restricted to secretarial or clerical work. Margaret Carter was one of a handful of employees hand-picked by the national airline in 1958. She and her contemporaries typified a new breed of post-war employees.

Pat Gregory Quilter had previously worked as a secretary in Sydney but had always wanted to travel. She attended an unnerving interview where she had to demonstrate deportment, make an on-board public address announcement and show a good grasp of current affairs. She got the job as a QANTAS flight hostess, initially on DC 4s.

There was only one hostess. QANTAS was a particularly male-dominated airline, and I think this was something to do with the very long-haul kind of flying that QANTAS did. We were probably one of the most long-haul structured airlines of any in the world. The hops were very long sectors because the planes were a lot slower. There was no movie, no video and no music to listen to. It was all very spartan with really only the view and the topics of interest we would explain as the captain pointed them out on the PA. We would be there to explain the history of the place and what it involved. We would chat to the passengers to keep them busy. We'd move around amongst them and talk to them quite a bit.

Vivienne Hanbury remembers the day she applied to join the airline, coming up to Sydney from Melbourne where she had worked for the Olympic Organising Committee in 1956.

I'd always wanted to fly. For various reasons I hadn't thought that I'd be able to. I thought you had to be rather nice-looking and have nice legs, and I didn't have nice legs so I didn't think I was too much cut out to fly in that regard. But that was silly. Anyhow I applied and I received an acceptance. I came up to Sydney and had a shower and changed in the Hyde Park ladies' rest-room there, and bought some Relaxatabs because I thought I'd be nervous, and I bought a hat and some gloves because I didn't wear a hat and I didn't wear gloves and I heard that really you should wear a hat and gloves; and toddled along to the interview.

I had the interview and there were four people there and they all asked me different questions. I had hair down to my shoulders then, and in those days you had to have your hair cut an inch above the collar. They said, 'Oh yes, you'll have to have your hair cut'. I said, 'Oh yes, I planned to have it cut above shoulder length'. They said, 'Oh no, no, no, you must have it an inch above your collar,' and I said, 'Oh well, I guess I could live with that'.

I realised that they would ask a lot of questions, so really I did a bit of homework as well as reading the sports page and the other pages that I was interested in—I did read the front page and brush up on my current events.

At that stage I think it was the Bank of New South Wales or the Commonwealth Bank had some very good pamphlets out on all the animals of Australia which were excellent

Top: The Lockheed Constellation. Bottom: A KLM Constellation interior

reference material. I went to QANTAS and got all the time-tables, really memorised all the places they went to, and had a good look at their routes and so on because I thought, well, there's no point in going to a company and not knowing where they go and what they do.

There were many duties hostesses were expected to perform.

The job that was the least glamorous was keeping the toilets tidy. In the early days that would seem to be a job that the hostesses had. We weren't supposed to look after the bar. We didn't have to actually prepare the meals, but we had a very definite part in serving the meals and one of our main things was to look after the mothers and babies and children on board, and to hand out magazines and things like that. We also had to do the paperwork which was quite extensive, for the crew as well as for the passengers. So we had a lot to do and people just seemed to have a false impression that we were just there as ornaments in some way. But in actual fact we had a lot to do.

When Margaret Carter joined QANTAS she had not imagined how tough it would be.

At times we crewed enormously long sectors, and if there were delays on those sectors our tours of duty could be twenty-two to twenty-four hours. That was not particularly unusual.

And there were sometimes unexpected duties for cabin crew.

I recall a flight back from South Africa. A Swedish oil tanker had blown up in the Indian Ocean, and they'd brought the bodies to the Cocos Islands, and we picked up both the living and the dead.

We took some of the bodies back to Perth. There weren't many first class passengers on that flight so we cleared the whole of the first class cabin and made that over to a sort of a mini hospital on board the aircraft. I learned one of flying's very interesting lessons then. I can recall the steward saying to me as they put these appallingly scarred burnt bodies on the aircraft, 'Well, Darls, they're all yours'. And that was always the attitude: if it came to care—mothers, children ailing, the aged, the young, the unwell, the dying, the dead—it was the hostess that took care. I didn't come from a trained nursing background and I can remember having to monitor saline drips and watch these men in the most really terrible condition, all this burnt flesh. If anyone's ever smelt human burnt flesh, and in the heat, it was just an unforgettable experience. But the up side of it was, I remember, those that we got to hospital in Perth later wrote to me and said, 'Thank you. You saved my life'.

Margaret Carter began her training on the last of the DC 4s flying up to New Guinea,

flying over the Owen Stanleys, without oxygen, of course. These were not pressurised aircraft. Sometimes because of the cloud levels we had to go so high that we had to watch the passengers for the blue in the nails and the passing out and there were many occasions in which I had walked the cabin with the oxygen under my arm, not least of all taking a puff myself every now and then.

Vivienne Hanbury also did her training on the Port Moresby run.

I flew all the way up there with twelve hours or whatever, flying through all these tropical storms and being delayed at either Brisbane or Port Moresby. Then we landed in Sydney. We taxied up to the door. I said to the girl that was training me, because I was supposed to stand at the door and say goodbye to the passengers. I said, 'I think you will have to say goodbye to the passengers because I'm going to be sick'.

Not, one hoped, over the smart uniforms QANTAS required of its hostesses, a wardrobe Pat Gregory Quilter can still count.

We had winter uniforms and summer uniforms. We had six white uniforms and a navy blue uniform. We had different shoes for each and the same cap, and of course when we left summer time Sydney, we'd get to London in winter, and you'd have to carry your big overcoat and you would take with you in cabin your six white summer uniforms, and before each port you were required to change into those so that before landing you were fresh and sparkling clean. You'd sit down and they used to be starched in those days, and they'd crease immediately, so it was a hard job looking bandbox fresh.

You'd be serving in the galley and you'd have to steady yourself while you were preparing, and many times we lost quite a few trays and food. The chief steward did all the meals in those days. The hostesses used to put parsley on top and carry the trays.

When former Manager of QANTAS cabin crew training, Dennis Liston, retired, ending his career supervising cabin service standards on 747s, he was an era away from the day he joined in 1951. In those days, after brief training, he was a flight attendant working on DC 4s.

I can remember we used to carry an umbrella in the galleys because they were never a pressurised aircraft, and I can tell you this, and it's a true story, that they used to leak in the rain. That was normal for them, I guess, and we used to have to put the umbrellas up, and when passengers would see us doing that it was just a strange sight to behold.

Flying, it seems, even in more modern post-war aircraft, retained some link with pre-war informality. Hostess, Daughne Kelpe worked with one captain who was particularly keen on fishing. Whenever they landed near a Pacific lagoon he would bring his fishing line.

During the hour's break, when they were re-fuelling the aircraft, he'd go fishing. He'd bring enough lines for about three or four of us to run over to the lagoon and fish. He'd insist on having fresh fish for breakfast.

We'd have to scale the fish and then let him know when he could pressurise the aircraft before take-off because we had to throw out the fish heads and the scales. So not only the crew, but also the passengers, had fresh fish for breakfast.

But Pacific or other ocean-crossing passengers wouldn't have fresh fish for much longer. The composition of the passenger list was shortly to change. Previously only a few could fly and aircraft could only fly a few. But Australia's post-war immigration policy brought a new kind of passenger and, in time, new aircraft.

Doug Worrad, former manager for KLM Australia, remembers an almost daily service in the peak years, 1949–1951, bringing migrants into Australia, mostly to Sydney, and occasionally to Melbourne. One of those migrant flights had an interesting ending.

There was a young lady whose fiance was in Australia and he

paid for her to come to Australia, and when they arrived in the Sydney terminal, she didn't want to come out of the Customs area, and we didn't understand this, but we found out that she had fallen in love with the purser on board. The Customs naturally insisted that she clear Customs and Quarantine and leave the place. So this poor fiance, he was standing outside with some flowers in his hand, waiting to welcome his coming bride, but we had to explain to the poor fellow that his fiancée didn't want to come out and she wanted to go straight back to Holland.

Being a charter flight, she couldn't travel back with that same aircraft because we're not allowed to carry any passengers on that, and then the crew each put in ten pounds and they bought her a ticket to fly from Sydney to Djakarta and there she got back on to this same aeroplane and flew back to Holland with the crew.

Some years later I was flying out to Holland and a purser came along to me and he said, 'You probably won't remember me, but you may remember the incident. Well, I'm the purser and I'd just like you to know that we got married, we have two children, and everything is beautiful.'

And in similar vein, Lennie Grey recalls that on one migrant flight, a captain had an unusual request from a couple of his passengers.

I don't know which country they came from but they urgently wanted him to marry them. They reckoned he was the captain of aircraft and he could do that. He couldn't convince them that he couldn't.

And there were more difficult situations, usually faced by the cabin crew. Pat Gregory Quilter handled plenty of language problems on flights through the Middle East.

It was a difficult experience because they couldn't speak the language and you couldn't get through to them. You had to try and explain to them. You had to explain to the Indians who were coming out that they couldn't light their little burners in the toilets and cook their rice.

We would leave Karachi at night, quite late I think it was, and I remember we had a full load of Pakistani seamen en route to Cairo, and you had to have all their documents done before you landed in Cairo and, of course, these men couldn't speak English and you couldn't speak Pakistani and you had to do all their documents yourself. When the lights were out in the aircraft and all they wanted to do was sleep, trying to get documents from their pockets was very difficult.

A difficulty compounded, Doug Worrad felt, by the sheer length of the flights.

In those days the aircraft would stop for a night stop in a hotel for the passengers in, say, Cairo and again in Karachi, again in Calcutta, again in Bangkok or Singapore, and then in Darwin. I often used to think of these migrants who had come from Holland, a country half the size of Tasmania. They'd been flying for four or five days. They would finally arrive in Australia in Darwin, and they must have thought, phew, we're there! But then after a night's sleep they would board the aircraft again the following morning, take off at seven o'clock and fly all day over the desert, hardly seeing another town, let alone any number of people anywhere, and arrive in Sydney at six o'clock that evening.

Newer, faster aircraft were needed. The Lockheed Constellation had been a QANTAS work-horse since the late 1940s but they were becoming less reliable. Margaret Carter worked on one or two latter day flights where mechanical problems were almost routine.

I've been on a flight to Singapore where two engines were cut and we were past the point of no return and so the flight continued on two engines in to Singapore, flying at a much lower level.

But what could replace the piston-engined Connie? In the 1950s the Constellation had dominated the international airline routes. America, largely because of the war, had overtaken Great Britain as an aircraft producer. But Britain was determined to regain its prestige in the air with a world first, a jetliner, the Comet.

Isolated Australia with some of the longest routes in the world was interested in the potential of the new jet. John Gunn recalls that in the early 1950s it looked as if, at last, Britain was going to leapfrog past the Americans with the first jet airliner.

It was a very exciting time. The promise for QANTAS and for BOAC was enormous. BOAC argued that they could dominate the airline world with the Comet, but again, Scotty Allan was one of the main minds in all these assessments, and Scotty Allan just said, no, the Comet wasn't the right aircraft for Australia.

There was a lot of pressure on QANTAS both from Britain and the Menzies government to buy the Comet. Scotty Allan was the key person in QANTAS management, advising and analysing and assessing all the aircraft available and he recommended against the purchase.

As chief test pilot, Allan was able to give an official reason, that the Comet did not have sufficient range to fly the Timor Sea, the gateway in and out of Australia. But he had other reservations.

I didn't trust its structure. The plating on the aeroplane was all lined up, long rows of it on the wings of the aeroplanes. Where the plates all conjoined, there were as many as nineteen plates all rivetted together, and of course they could rip open if it broke at that place—and it did.

The tragic outcome—a succession of crashes and the eventual grounding of the Comet—is now part of aviation history, the price Britain paid for its pioneering of the world's first jet-liner. There were suggestions that the Comet was politically flown, with insufficient time to test its new technology.

Fortunately for Australia, there were many other reasons not to buy it. In the judgment of senior QANTAS staff the early Comet lacked range, it was under-powered and lacked accommodation for both passengers and freight. It was lucky for QANTAS that they had not entered any purchase agreement with Britain.

The Comet was to emerge as a successful airliner in its later versions but Australia couldn't afford to wait for the costly solution to the metal fatigue which had caused the early Comets to fall out of the sky. Scotty Allan recalls that

the British High Commissioner was somewhat disgusted because we decided not to fly the Comet. At that time the Comet was the only alternative aeroplane. In due course

I was sent to America to look for an aeroplane to take its place.

QANTAS staff went to the Boeing factory in Seattle to evaluate the rival which became the Comet's successor, the Boeing 707. At the time there were no real alternative aircraft if Australia was not to be left behind in international aviation. The plane was to prove a success with both passengers and crew from the outset.

Cabin steward Dennis Liston came back to Sydney after flying in New Guinea,

> riding around on DC 3s for two years, and my first trip of course, was on a 707. I didn't quite believe it because I ended up down the back of the bus as we called it, in the last jump seat, and this thing took off. I became the white knuckle brigade on that first flight. It was a night time flight up to Singapore or Hong Kong. We thundered down the runway and I was in the very back seat because it was my first trip on it, and there was no braking effect from the engines, so you just kept going.

In 1959 Marie Heald was on duty in the Sydney QANTAS switchboard the day the first Boeing was expected to arrive over Sydney.

> It was lunchtime and I was there on my own because the traffic wasn't so heavy during the lunch hour. We had ever so many calls from people wanting to see this Boeing which was supposed to circle the city three times before it landed, and one man rang in from Marrickville, and he complained. He said, 'That Boring,' (he called it a Boring instead of a Boeing), 'It's supposed to fly round the city three times. I haven't seen it and I'm a taxpayer and I'm entitled to see it. I've helped pay for that!'
>
> I had hoped to go on holidays a few months after I transferred to the head office, hoping to go on this new aircraft, but it was booked out—fully booked out. At that time I didn't think it would be so well-received and that it would be difficult to get on a flight, but people were so keen to fly on this lovely new aircraft.

For hard-headed airline executives the economic effect benefit of the jet airliner was its biggest advantage. Payload and efficiency improved. The jet engine was to prove more reliable with lower maintenance costs. As QANTAS engineer Ernest Aldis observed, it was to prove the biggest technological breakthrough in aviation.

Absolutely unseen, unheard of, and unthinkable at that time. From World War One onwards they were battling with the piston engines right up to the Super Constellation aircraft which was the last of the piston engines. Millions and millions of dollars were spent on the research and development of the exhaust valve in the piston engines at that time; but they never did solve the problem of valve failure prior to introduction of the gas turbine engine.

Jim Cowan remembers that when QANTAS first took on 707s

they were flying the Pacific, and one of my friends who was in Honolulu said, 'Jim, the great difference between Connie and the 707 is that when the 707 leaves, you can go home and go to bed. It won't come back.'

The sheer size of the 707 made a great impression on cabin crew. Vivienne Hanbury appreciated the extra space.

Of course everyone thought they were like big cigars. We thought they were huge—monstrous. There were more stewards but not any more hostesses when we first started flying on the 707s. There was only one flight hostess and we were supposed to work at the first class cabin and the economy class cabin, and at least try and go through. Sometimes if you had a full first class cabin and it was a short flight, it would be all you could do to get down through the economy class cabin and try and give out magazines, or attend to babies and mothers. On a longer flight, of course, you finished your first class meal and then you could attend to the economy, because you had to physically take part in a meal service. Then you could go down and look after the economy classes as far as magazines and blankets and so on.

But there were compensations, as Margaret Carter came to appreciate.

These aircraft offered enormous comfort compared with the Super Constellations by virtue of the speed with which they got you there. Speed became the buzz word. Often the sectors that we crewed altered because we weren't working such long tours of duty. The flying time on piston engines compared with jets—I mean it was almost half, and of course, we didn't stay so long in the ports where the crews slept. In the Super Connies we would often have five days, say, in Singapore or New York. On a trip to London via New York, it would be three days one way and two the next, with about three or four days in London.

For Dennis Liston the biggest change was the increase in cabin staff. There were now three crew working in first class and another three in economy, depending on the aircraft configuration. But, as he recalls, with the introduction of the 707, crews saw their homes and families more frequently.

In the old Connie days you could be away for just about a month because of the frequency. With the advent of the 707 the times were chopped down practically by half. But in the Connie days one would get a phone call to say you were going off somewhere and that was it. You were off for two or three weeks at a time and came back not knowing where you were going next trip, and that was pretty dramatic on family life. In the case of people with families, every time you went away that was when the children were sick, or that was when the emergency happened or that was when the plumbing broke down or all those sorts of things. And unfortunately it's had its toll on a lot of people.

Vivienne Hanbury felt that early post-war flying life strained a lot of marriages.

Some of the wives could handle it but some of them couldn't. They had to be mother and father while their husbands were away. There were opportunities for people to play up while they were away and some did and some didn't, and some of the ones who did were wonderful husbands when they were at home. Some of the ones who didn't, when they came home never took their wives anywhere because when they were away they were out. Some of our aircrew, when they came home, all they wanted to do was stay home; they didn't want to go anywhere. Their wives had been at home and they wanted to go out and have a good time. The husbands had had to eat out every night when they were

Top: Daughne Kelpe photographed in Egypt in 1961
Bottom: QANTAS's first jetliner, the Boeing 707

away, and they'd usually gone out even if they didn't usually go out when they were at home. They'd go to movies or dancing or to shows or whatever. So they'd had quite a bit of nightlife and a lot of them just wanted to stay home when they got home.

The 707s not only reduced time in the air, they also provided the first step towards mass travel. Former KLM Australia manager, Doug Worrad observed another interesting development.

Without the migrant population in Australia, civil aviation from Australia would not have made anything like the progress that it has made. It was very largely dominated by the movements of the migrant population. In 1964 there was an explosion of passenger traffic by air. We used to say that that was because the migrant population, certainly the Dutch, and I could imagine other nationalities, too, had been in Australia approximately fifteen years by that time from, say, 1953 to the mid-1960s. They had organised their housing. They had their motor cars, and they'd paid for their refrigerator, and then all of a sudden they were financially able to consider a flight to their original homeland. At the same time the shipping companies virtually ceased passenger transportation between Australia and Europe. Then the airlines started introducing excursion fares and so on, and all of a sudden it all came together and it was possible for very large numbers of passengers to consider flying over to Europe.

The effect on migrants was inevitably to encourage a greater sense of identity with Australia. They knew now that they could be back in Europe within a day or so to see home and elderly relatives. But they could live and work in Australia.

With larger numbers now travelling, the old patterns of relaxation and conversation between passengers and crew began to disappear. Skipper Phil Mathiesen recalls that until the end of the Connie days

it was possible for the captain to liaise with the passengers, and whilst not as much as with the flying-boat days where you really had the chance with the lesser number and longer flights, with the Connie you could still do that. But came the 707 and it became impossible to really take much interest in the passengers.

Other changes were taking place, too, at the helm of the national airline—probably the biggest since 1919. Towards the end of 1965, Hudson Fysh, who was one of the three founders of the airline and the last living link with the beginning of QANTAS, was nearing the end of his term. He had seen unimaginable technological changes in his working lifetime, from flying bi-planes in World War One to managing an airline that flew jets half-way round the world each day.

Later, in retirement, he suggested that QANTAS was not ready for yet another big change. The 707 and the Constellation had seemed big enough changes at the time but, in his words, they ensured a period of steady progress.

> Now on top of this have come the 'Jumbo' jets, costing some twenty six million dollars a piece, and carrying up to 490 passengers, a tremendous single leap ahead, and bringing down the cost per seat mile—that is, the cost of a seat as it flies empty. Ideal for the North Atlantic high density trade but as yet, I feel, too large for the Australian business without unwarranted sacrifices in frequency. QANTAS, in my personal view, has been obliged to acquire the type perhaps earlier than it should have, in order to compete with Pan-American Airways who are operating the Jumbo to Australia.

QANTAS now operates an all Boeing fleet of 747 and 767s. A 777 with folding wings is on the drawing board and further down the track is the 800-seater Super-Jumbo, which may or may not be built.

The radio series *Rag Sticks & Wire*, on which this book is largely based, was launched in the forward cabin of the new 747-400. It easily held many of the contributors including former aircrew and passengers, and journalists. At that enjoyable ceremony John White of QANTAS reminded me that the length of the fuselage we were sitting in was longer than the distance covered by Wilbur and Orville Wright's first powered flight in 1903.

Technology apart, it is interesting to observe human reaction to change. Hudson Fysh may have had commercial reservations about the wisdom of buying the 747, but not his cabin staff. As Dennis Liston cheerfully admits,

cabin management was totally sold on the 747. I mean the 747 was the greatest thing since sliced bread. We loved the aeroplane but the crew didn't at first. They loved the little aeroplane, the 707.

They liked the closeness of the cabins and the small crew, but then once they started flying on the thing and getting used to it—I mean it was a bit of a disaster from our point of view in the early days but we soon rectified that. Everything was going wrong, there were teething problems, and I remember clearly, I think it was the first trip I did, the guy that was working in the galley got so frustrated, he opened the oven and the door came off in his hands. Those sorts of things happened. The numbers all of a sudden increased— one day we had 100 passengers, and the next day were on with 300. It was a bit of a shock.

Cabin steward Lorenzo Montesini saw the phasing out of the 707s and their replacement by the Jumbo as a major change.

They had been the mainstay of the fleet for twenty years, 1959 till 1979, and the period I joined was a very exciting new time when the carrier went entirely for wide-bodied jets. It was just post-Whitlam time and I think the carrier was changing its image of itself; Australia's image of itself was changing. The image of QANTAS cabin crew as the tennis-playing, blond, tall, that sort of look, thank God, had been put to rest once and for all, and they were quite happy to have people who were more interesting, had more to bring, more variety to bring as cabin crew, and that's what made it a very interesting time.

But he has some regrets about the new technology.

The bigger the aircraft, the more you lose a sensation of flying—even the mere sensation of walking across a tarmac to go to an aircraft to see the physical struts, and the wheels, walking into the aircraft, and the smell—the mere smell of kerosene in the air, that is a concept in travel perhaps which has changed. Now you're just going from a room to another room.

The passengers have become much more blase now even

in the ten years that I've been flying. We occasionally go and say to a passenger, 'Would you care to go up to the cockpit?', and they say, 'oh no'. I find that extraordinary.

In 1990 when QANTAS acquired the latest version of the Jumbo, the 747-400 built to fly even further and faster, they appropriately called their new 747 fleet, 'Longreach'. The name also honours the small far west Queensland town where QANTAS was first based.

Some voices from Longreach recall an Australia when aeroplanes were little more than rags, sticks and wire. Mary Yeoman was possibly QANTAS's first paying passenger in 1919.

Hudson Fysh and his friend bought the old plane after they came home from World War One. We had a lovely flight around and I thoroughly enjoyed it, and of course he said to everybody, 'How did you enjoy it?', and I said, 'Oh, I loved it'. I said, 'If I ever live long enough and have enough money, I'd love to fly round the world'. Well, I have done just that.

Edwin Chapman's father had dreams of the future.

He said aeroplanes were going to improve tremendously, and I just couldn't envisage it. I never imagined anything like a 747.

In 1921, as a small girl, Nancy Button made her first flight with QANTAS co-founder Paul McGinness. Strapped into his tiny bi-plane, she flew from her father's property a few miles to land at the new Longreach aerodrome.

In 1988, with her husband and friends from Longreach, she went to England in a QANTAS 747.

Sitting there with all those passengers and that big plane, I thought of the early planes out here and the open cockpits and just absolutely marvelled at them, how they had improved from early days. I'm quite sure that Fysh and McGinness and Michod and McMaster could never foresee what would come.